FLAT EARTH

**A Plane Truth from
the Ancients
to the Abrahamic Bibles
to Modern Times**

James W. Lee
Iconoclast

Copyright © 2017 by James W. Lee
Flat Earth; Investigations into a Massive 500-Year Heliocentric Lie
First Edition
All right preserved. No part of this book may be reproduced in any form or by any electronic or mechanical means including information storage and retrieval systems without written permission from the author. Reviews may quote brief passages to be printed in a magazine or newspaper.

ISBN-10:
1542805333
ISBN-13:
978-1542805339

Layout, design, editing:
Ellen Sklar-Abbott
Lenora Henson
Melanie Moran

Front Cover: THE FLAMMARION engraving of 1888 beautifully illustrates medieval cosmology, featuring a flat earth bounded by a solid and opaque sky, or firmament. The Greek astronomers represented this 'ceiling' as formed of a solid crystal substance; and prior to Copernicus, a large number of astronomers thought it was as solid as plate-glass.

The engraving, which represents the scientific or the mystical quests for greater knowledge, depicts a man clothed in a long robe and carrying a staff, kneeling down to poke his head through a gap in the star-studded firmament, thereby discovering the celestial workings of the greater reality beyond.

Flammarion's accompanying caption reads: "A missionary of the Middle Ages tells that he had found the point where the sky and the Earth touch.

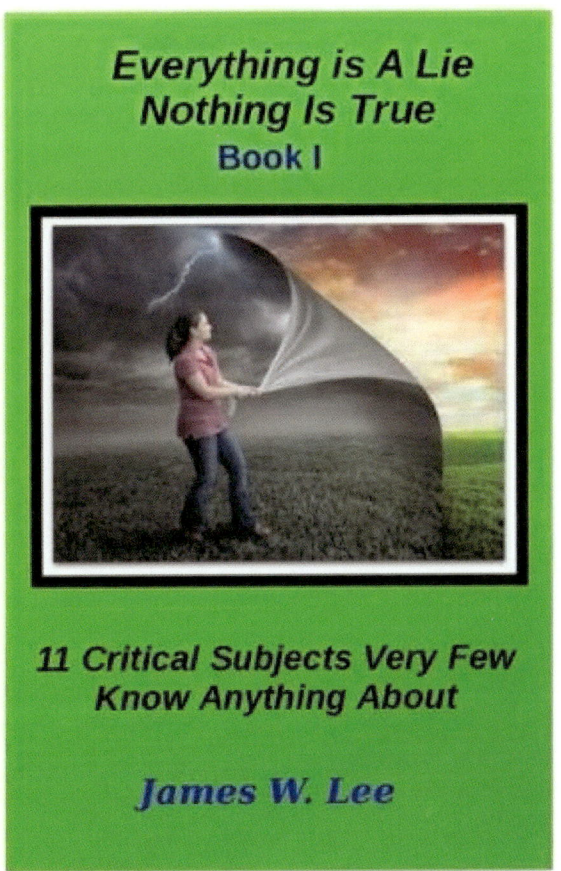

Order books:
Color copy
https://www.createspace.com/6890596
Black & White copy
https://www.createspace.com/6896622

Order books:
Black and White copy
https://www.createspace.com/7003126

Websites
www.tabublog.com
www.aplanetruth.info
www.avvi.info

You Tube Channel:
aplanetruth.info

All Health Store
TheDWDGshop
(Summer 2017)

Dedication

This book is dedicated to those who have had such a deep and meaningful impact in my life.

To Barbara, Herb and Jan Lee. Thank you for your many selfless examples of actively giving back to help heal and care for all. Thank you.

To all my inspirational and motivational guides and teacher; Master Robert Ayers. the David &Terry Lee Family, the Juan Malfavon Family, the Wendell Rand Family, the Tim Ward Family. Bill and Susan Lee, DK & KK, the Bart Adams Family, Geognostic Bruce, Jane "Rudy" Lorand, Yoga Bill & Soul Force Lisa. Grant Grinnell & Divine Melanie Moran. Bless you all and thank you.

I also dedicate this body of work to all creatures of Nature. May we immediately find ways to stop treating our only Sacred Mother and Father Sky as solely a sewer, a supply house, and a very sick lab experiment. Nature Bats Last is the lesson she will be teaching us in the years ahead.

To my son, Jaxson, you have taught me what unconditional Love is, I could not ask for more. This book is dedicated to the education of all youth of today. May all think for themselves and always seek the truth. wherever it may lead.

With profound gratitude,

Jamie Lee
March 11, 2017

Prologue

Cassandra's Dilemma

The ancient Greek mythology entitled "Cassandra's Dilemma" tells the story of a mere mortal, Cassandra, the most beautiful woman ever to have lived in the times when the Gods interacted directly with humans on the physical plane.

Apollo, the most powerful and most handsome of the sons of Zeus, took one look at Cassandra and fell immediately in love. He then set about to marry her. It would be the first ever marriage between a mortal and a God. When Cassandra's girlfriends learned that Apollo was in love with her, they envied her greatly. "Oh, he is the most divine of the Gods," her girlfriends cooed. "You are the most fortunate of all mortals to have a true God in love with you and wish to marry."

Yet during the initial courtship, Cassandra would have nothing to do with him. Cassandra was not impressed with Apollo and his God-sized ego. "He's soooo arrogant, he's cocky and an incredible narcissist," she reported to her girlfriends the day after. "All he talks about is how great he is and how lucky I am for him to be in love with me," she recalled. "And besides, I only marry for love and I could never love someone as self-centered, arrogant, cocky and conceited as Apollo!" Upon learning of Cassandra's comments, Apollo flew into rage. No women, goddess or mortal had ever denied him like this before. "Who is she to defy the love of Apollo?" He spoke aloud to the heavens. "She has no idea what honor I am bestowing upon her with my love and affection," he proclaimed. "I will win her love or she will suffer my wrath for such indolence and disrespect!" he announced to all.

Back in ancient Greek times, Gods could take human form. So, on Apollo's next visit to Earth he had a surprise for Cassandra. "Cassandra," Apollo began in a soft, loving tone, "You are the most beautiful of all women that Greece has ever born and I choose you as my wife to rule at my side in the heavens." He then leaned over and gave her a long, passionate kiss. With that kiss, he breathed into her lungs with his breath the Gift of Prophecy, the non-mortal ability to see into the future, that only Gods possessed.

With Cassandra's new visionary power, she could win every time at the local chariot races and became very wealthy. She counselled her girlfriends as to who were appropriate suitors as well as prevent injury and illness to many in her village of Troy. In short time, she fell for Apollo and agreed to be his bride. The wedding was set and it was to be the grandest of all weddings that had ever been since no God had ever set to marry a mere mortal before.

The day of the wedding was like no other. The Pantheon was laid out in glorious rose bouquets, while the highest priest of all Gods, Zeus, Apollo's father would marry them at the altar. Neptune, the God of Ocean, Music, and Wisdom would lead the musical ensemble while Demeter tended to the fresh, garden food that would be served. The big day arrived and Apollo took his place on the right hand of his father, Zeus, on the altar.

The music began as the bridesmaids lead Cassandra down the processional aisle, yet just as she left her vestibule, she stopped and turned to her maid of honor and declared simply. "I can't marry him!" She continued, "I promised myself that I only will marry for true love and in my heart of hearts I truly do not love Apollo.

Please go tell him that there will be no wedding." She then turned quickly and returned to her bridal room alone. Her Maid of Honor sadly delivered the devastating news to Apollo in front of all the other Gods and mere mortals and all Hell broke loose. Upon hearing the news, Apollo flew into a rage for the Ages, blowing gusts of winds that destroyed the wedding altar and in his exalted fury, brought down bolts of lightning and booming sounds of angry thunder. He then descended the vestibule where Cassandra sat to confront the only mortal ever to deny a God in such an embarrassing manner. As he approached, Cassandra sat stoically and calm, assured she had made the right decision. "How dare you embarrass me in front of all the Gods, you mere mortal!" He shouted at her. "You have no right to deny me."

"Oh, I see," Cassandra calmly replied, "You are more worried about saving face with your Gods than your love for me. She took a deep breath and measured here next words carefully. "Now I know you have just confirmed that I have made the right decision not to marry you, Apollo. It's over." Apollo then tried everything he could to change Cassandra's mind, yet finally came to realize that he had lost her for good and there would never be a wedding.

In desperation, he made one last plea to her, "Cassandra, give me one last kiss, and I will be out of your life forevermore." "I'll do it," she retorted in exasperation, "I'll do anything just to get you to leave me alone." Apollo leaned in and gave her one last kiss. As their lips touched, he breathed into her one last spell, that no matter what she said going forward with her powers of prophecy, no one would believe a word she was saying for the rest of her days. Cassandra had to live out the rest of her life knowing her parents and her home of Troy was going to be overrun and all killed by King Agamemnon and his Greek Army as she tried in desperation and vain to halt the slaughter of her loved ones.

She lived the rest of her full life in guilt and shame for not being able to save her family, friends and community from death and disaster.

This is known in mythology as *"Cassandra's Dilemma."*

Table of Contents

Dedication 5

Prologue 6

Introduction 11

Chapter 1 - 5,000 Years of Flat Earth TheoCosmology 39

Chapter 2 - Flat Earth Bible and Qu'ran Citation 49

Chapter 3 - Flat Earth Is 63

Chapter 4 - A Flat Earth Cons-piracy Case 103

Chapter 5 - Flat Earth Gallery & Quotes 145

Epilogue 156

Appendix I - The Rulers Who Make the Rules 161

Appendix II - Secret Societies That Run the World 165

Bibliography/Resources 175

"It may be boldly asked where can the man be found, possessing the extraordinary gifts of Newton, who could suffer himself to be deluded by such a hocus-pocus, if he had not in the first instance willfully deceived himself? Only those who know the strength of self-deception, and the extent to which it sometimes trenches on dishonesty, are in a condition to explain the conduct of Newton and of Newton's school.

To support his unnatural theory Newton heaps fiction upon fiction, seeking to dazzle where he cannot convince. In whatever way or manner may have occurred this business, I must still say that I curse this modern theory of Cosmogony, and hope that perchance there may appear, in due time, some young scientist of genius, who will pick up courage enough to upset this universally disseminated delirium of lunatics.

Someday someone will write a pathology of experimental physics and bring to light all those swindles which subvert our reason, beguile our judgement and, what is worse, stand in the way of any practical progress. The phenomena must be freed once and for all from their grim torture chamber of empiricism, mechanism, and dogmatism; they must be brought before the jury of man's common sense. "

~ Johann Wolfgang Von Goethe ~ German Poet and Philosopher (1749-1832)

Introduction

Alternate Realities

This book is like no other book you have ever read before.

For those with open minds and an ability to think for one's self, the information contained within can be profoundly liberating to your conscious mind. You may discover a true reality of all that surrounds us in the physical realms of our common existence to be a completely different narrative from what you were taught in school since birth. You may find a reconnection and remembrances to a very rich and ancient ancestry of a much greater story about who we are, where we came from, and where we go when our Spirit returns home.

This book may be highly disturbing as one begins to comprehend that the round ball heliocentric story of modern science is likely the Greatest Lie in all modern history. It can be deeply upsetting to learn that NASA controls all news from a fake space and that no human has ever physically traveled outside of our domed home. You may feel awed in learning that you now possess more knowledge about our real world than 7 billion + others on our non-planet.

The Great Lie has been told and sold over and over again to billions and billions of people, over hundreds of years, and never questioned until now. All ingrained and indoctrinated since birth through imagery, illusion and dark magic. A Great Lie expertly crafted and installed in our minds through social media and fictional historical narratives where to even question the official narrative of modern heliocentric theoretical science deems you crazy by a compliant and obedient society. It may also be very upsetting to learn behind it all is a very dark Luciferic energy of Men in Black. Secret society and religious zealots who work in the shadows to keep our souls from truly knowing who we truly are and our place in the Cosmos. The same ones who have openly declared as their 4th Sacred Vow to "banish all heretics and Christians from the face of this Earth." (Note: This is nearly the same exact language 45th US President, Donald Trump, declared just moments into his 2017 Presidential inauguration acceptance speech, heard by hundreds of millions of people, prominently promising to eradicate the threat of "radical Islamic terrorism from the face of the Earth." In the following days, he successfully lobbied for the US Congress to add tens of billions more to the US $1 Trillion budget per year. Once a religious Holy War is declared and targeted, all religions become targets, even Christianity!)

As we learn and awaken to Flat Earth cosmology, many are finding a much more authentic, rich and intimate personal understanding of the Story of Us. So important that all the Cosmos is centered around Earth itself and of the greatest importance in all of Creation. A new awakening that aligns conclusively with the Abrahamic bibles. That reconnects profoundly to our common past.

> **Pope Boniface VIII**
> We declare, say, define, and pronounce that it is absolutely necessary for the salvation of every human creature to be subject to the Roman Pontiff.
> — AZQUOTES

That is of such profound grand design and beauty that we just now begin to realize that the Sun, the Moon, and The Luminaires are all set in the sky to assist humanity to achieve their own greatest potential.

"The Modern Skeptic: The Greek word 'skepsis' means investigation. By calling themselves skeptics, the ancient skeptics thus described themselves as investigators. They also called themselves 'those who suspend', thereby signaling that their investigations lead them to suspension of judgment. They do not put forward theories, and they do not deny that knowledge can be found.

At its core, ancient skepticism is a way of life devoted to inquiry. It is as much concerned with belief as with knowledge. As long as knowledge has not been attained, the skeptics aim not to affirm anything. This gives rise to their most controversial ambition: 'a life without belief'. Skeptic does not mean him who doubts, but him who investigates or researches as opposed to him who asserts and thinks that he has found." [Miguel de Unamuno, "Essays and Soliloquies," 1924]

"Children are taught in their geography books, when too young to apprehend aright the meaning of such things, that the world is a great globe revolving around the sun, and the story is repeated continuously, year by year, till they reach maturity, at which time they generally become so absorbed in other matters as to be indifferent as to whether the teaching be true or not, and, as they hear of nobody contradicting it, they presume that it must be the correct thing, if not to believe at least to receive it as a fact.

"They thus tacitly give their assent to a theory which, if it had first been presented to them at what are called 'years of discretion,' they would at once have rejected. The consequences of evil-teaching, whether in religion or in science, are far more disastrous than is generally supposed, especially in a luxurious laissez-faire age like our own. The intellect becomes weakened and the conscience seared." —David Wardlaw Scott, Terra Firma: The Earth not a Planet Proved from Scripture, Reason, and Fact

How can it be that for over 5,000 years, as far back as written modern history allows us, that nearly every single culture on Earth believed the land we have all lived upon since our birth was known to be flat, motionless, and the center of all this Creation?

Nearly all ancient cultures, previous to the current Roman Dynasty (310 AD–present), taught Flat Earth Theocosmology. The Ancient Vedic's of India, one of the first civilizations to write their thoughts and ideas down, held firm that this one Creation, of which there are many, lasts for some 311 trillion, 40 billion years of a single Manvantara. A Manvantara is just one Brahma creation and destruction of a universe, ruled over by Manu. Compare that to our modern history and "pre-history" of only some ten thousand years and you get an idea that humanity is much greater in scope and design than just a few millenniums of time. Why have such rich ancient Flat Earth teachings from so many cultures been omitted from western education and academic discussion?

From the Ancient Chinese, to the Hindus, Buddhists, Sumerians, Chaldeans, Babylonians, Egyptians, Ancient Hebrews, Greeks, and Nordic Vikings, all held that we were the center of this Universe, inside a vaulted roof or dome, surrounded by water above and water below. Additionally, the Mayans, Toltec's, Aztecs, Cherokees and many Native American tribes shared similar Theocosmology holding great reverence to Mother Earth and Father Sky.

You were not taught in schools the fact that The Big Bang theory was not proven by Albert Einstein but was actually written by a Jesuit priest, Father Georges Lamaitre who is said to have based his work on Einstein's theories. Why would a Roman Catholic priest write the Big Bang Theory when the Catholic Church was supposedly proven in error as to the Earth's shape and motion by Galileo, Copernicus, Newton, et. al?

It wasn't until the mid-16th that the Vatican was said to be proven wrong about the shape of the Earth by scientific proof. Proof through technology and mathematics that we were moving around the Sun in an orbit and moving and Earth was a rapidly spinning round ball sphere. The early astronomers, from Copernicus to Newton to Albert Einstein all had intimate Roman Catholic connections. Nicolaus Copernicus dedicated his book to Pope Paul III and died a priest. Sir Isaac Newton's 4th edition book, Principia, (that described gravity for the first time), was final edited by Jesuits, also known as the Society of Jesus, the most powerful branch of the Roman Catholic Vatican.

All have been told and sold that it was the Vatican who was proven wrong when friend of the Jesuits, Catholic Galileo Galilee started grinding his own glass to make telescopes so he could observe the movement of the stars through magnification. He concluded that the Earth was in motion relative to the Sun (heliocentric) and not the other way around (geocentric). Galileo claimed to observe that the Sun was not just the fixed center of the solar system but the fixed center of the entire universe. Cardinal Robert Bellarmine, a Jesuit—one of the most important Catholic theologians of the day—issued a certificate that, although it forbade Galileo to hold or defend the heliocentric theory, did not prevent him from conjecturing it.

The heliocentric story builds from there that the Roman Catholic Church was proven wrong about the motions of the luminaries above and their relation to Earth, effectively discarding 5,000 years of geocentric (Earth centered) Theocosmology due to scientific "proof". In 1992, some 350 years after the Roman Church's condemnation of Galileo, the Vatican publicly admitted that "we today know that Galileo was right in adopting the Copernican astronomical theory" spoken by Paul Cardinal Poupard, at the formal closing of a 13-year investigation into the Church's condemnation of Galileo in 1633. (The Vatican is the largest and longest owners of telescopes and observatories in the world with the most intellectual property patents. That it took the Church 350 years to admit its greatest mistake in Theocosmology is a big red flag as to the veracity of the history of origins of heliocentric proof and scientific fact.)

The same empirical evidence that Galileo used as "proof" of Earth orbiting the Sun can be debunked simply by personal self-observation. At the Sun's rising, note the position of the Sun on the Eastern horizon. Over the course of the next minutes and hours watch as the Sun *moves* across the sky and then *sets* in the East. You do not need a telescope to see the Sun *move!* You have just disproved all of heliocentric theory with one simple self-observation.

When the heliocentric story was originally being sold for mass indoctrination in the 1850's, those wishing to sell the heliocentric theory to the masses could not convince those that lived outdoors and worked outdoors without artificial light pollution that the Sun was not moving. This may be to this very day why we still call the celestial *movement* of our Sun, a "Sunrise" and "Sunset". (Why haven't Astronomers, Academicians, English teachers, et. al., ever corrected such a gross misuse of basic grammatical language errors through modern history, if the Sun is still and unmoving relative to Earth? What is the correct heliocentric terminology? An "Earthrise" and an "Earthset"?....can you say awkward?!?

"Thus, in 1600, there was no official Catholic position on the Copernican system and it was certainly not a heresy. Heresy literally means 'choice' or 'choosing' for themselves what to believe and practice was considered a crime against the Roman Catholic Church, whose priests developed 'The Inquisition' to discourage further stray from the church's established dogma of suppression of basic human rights and needs. Very little has changed today" ~ William 'Dean' Garner, from "Mediaeval Heresy and the Inquisition" by Arthur S. Tuberville

After the Romans pillaged and burned the Great Libraries of Alexandria by Roman Emperor Julius Caesar around 48 B.C., they have controlled most of our modern history. They began to tell the history they wished for us all to learn, not the true origins of humans, as told by ancient civilizations. History is told by the winners, and no one civilization or religion has ever been more successful in their conquests in history than the Holy Roman Empire. When the Guttenberg book press (circa 1440) was first invented, the largest owner of books published, and book publishing companies thereafter, was the Roman Catholic Church. Many times, they used alias' and ghost writers to hide the true authors of history books used in schools and translated in many languages.

It was called the Dark Ages (500-?) because no light of alternative knowledge was allowed to escape control of the Vatican teachings or you were declared a heretic. The Vatican used the new printing presses and the translation of historical narrative into other languages than Latin, to write history as they wished. The Ages of Renaissance (1300-1600) and Enlightenment (1685-1815) were said to be an awakening of consciousness in the minds of the many from the religious few. Science triumphed over religion was proof of fact. The Vatican was also the first to translate books from Latin and ancient Greek into other languages. Like religion knows so well, "get by six, you got 'em for life". Roman Catholics version of our common history, once sold to a generation, is accepted as fact by subsequent generations.

Why, in the mid-1500's, were the first astronomers to travel around the world to China and India and Jesuits? Why did the Jesuits bring over the very first celestial observatory to the United States in the early 1600's if they were proven so wrong about our common cosmology by Galileo and Copernicus?

Many ancient cultures taught our life mythology was played out in the stars above if we knew how to interpret their meanings. This was the ancient's classroom under the luminaries as most cultures lived and slept under the stars each night. The ancients, before modern science technology, thousands of years ago, accurately predicted eclipses and comets. They built huge stone pyramids in perfect alignment with the stars above.

Today, we have lost our connection to the heavenly classroom through the loss of oral tradition, Industrial (D)evolution and light pollution. Rather than our common story of us being just one of billions and billions of planets, in a small insignificant corner of our universe, geocentrism teaches that we are the epi-center of all this creation.

"Seeing is believing?" Another simple observation you can do is to lay out at night and focus on the North "Pole" star. You will see over time, other star's circle around the fixed pole star, thereby easily proving once again, we are stationary and unmoving relative to the luminaries above.

One example of ancient culture's handiwork relating to the heavens above are the huge and vast *Nazca Lines in Peru* that can only be seen in full from high up in the air. These lines, still etched into the Earth today, were carved out thousands of years-ago. The ancients left their teachings and legends carved into stone and lands as permanent record for all to learn from. Their hieroglyphs, carved onto obelisks and on pyramids are said to tell up to ten thousand words per glyph, or image. The ancient Mayan calendar, written in stone over 5,125 years ago, correctly predicted the transition of the Earth's movement through the Milky Way, or Dark Rift. These calculations correspond exactly with ancient Greek Plato's Great Year. Plato's Great Year cycle covered one full procession of equinox or 25,920 years-time. To better understand how advanced ancient record keeping was over modern technology, consider that we still have their records over thousands of years-time while modern Man records its history today on memory discs and thumb drives!

The Great Pyramid of Giza in Egypt is the most accurately aligned structure in existence and faces true north with only 3/60th of a degree of error. The position of the North Pole moves over time and the pyramid was exactly aligned at one time. The Great Pyramid sits upon 13 acres of land fill. After thousands of years, the angle of tilt from level plane at the base of the Great Pyramid is less than a few inches on any one side. The interior temperature is constant and equals the average temperature of the earth, 20 Degrees Celsius. It is located at the center of the land mass of the earth. The east/west parallel

that crosses the most land and the north/south meridian that crosses the most land intersect in two places on the earth, one in the ocean and the other at the Great Pyramid. The outer mantle was composed of 144,000 casing stones, all of them highly polished and flat to an accuracy of 1/100th of an inch, about 100 inches thick and weighing approx. 15 tons each.

The centers of the four sides are indented with an extraordinary degree of precision forming the only 8-sided pyramid; this effect is not visible from the ground or from a distance but only from the air, and then only under the proper lighting conditions. This phenomenon is only detectable from the air at dawn and sunset on the spring and autumn equinoxes, when the sun casts shadows on the pyramid. The granite coffer in the "King's Chamber" is too big to fit through the passages and so it must have been put in place during construction. The coffer was made of a block of solid granite. This would have required bronze saws 8-9 ft. long set with teeth of sapphires. Hollowing out of the interior would require tubular drills of the same material applied with a tremendous vertical force. Microscopic analysis of the coffer reveals that it was made with a fixed point drill that used hard jewel bits and a drilling force of 2 tons.

The cornerstone foundations of the pyramid have ball and socket construction capable of dealing with heat expansion and earthquakes. The mortar used is of an unknown origin. It has been analyzed, and its chemical composition is known, but it can't be reproduced. It is stronger than the stone and still holding up today. It was originally covered with casing stones (made of highly polished limestone). These casing stones reflected the sun's light and made the pyramid shine like a jewel. It has been calculated that the original pyramid with its casing stones would act like gigantic mirrors and reflect light so powerful that it would be visible from the moon as a shining star on earth. Appropriately, the ancient Egyptians called the Great Pyramid "Ikhet", meaning the "Glorious Light". How these blocks were transported and assembled into the pyramid is still a mystery.

Some of these granite stones were transported from Aswan, Egypt 500 miles away. It has been estimated that they cut, carved, erected and set in place around 8,000 metric tons of granite, 6 million metric tons of limestone, and half a million metric tons of mortar in the Great Pyramid alone in what modern history estimates to be entirely constructed in 40 years-time!

The three pyramids at Giza line up perfectly with the Constellation of Orion's belt in the sky. They purposely built these monuments to be in concert with the luminaries, or stars above. The Giza pyramids are now thought to have been built some 10,000 years-ago when the constellation Leo was directly opposite the Great Sphinx. This cannot be by coincidence but by intelligent design. Modern science does not understand, nor can they duplicate, nor do we to this day understand even why they built such massive engineering feats. These Egyptians, of such magnificent engineering and Theocosmology, were of geocentric Flat Earth faith.

Astrotheology is the study of the astronomical origins of religion; how gods, goddesses, and demons are personifications of astronomical phenomena such as lunar eclipses, planetary alignments, and apparent interactions of planetary bodies with stars. ~ Wikipedia

Our distant ancestors intimately followed the movement of the Sun, Moon, and stars because their lives depended on it. Springtime was a time of celebration and joy because the Sun had risen enough so seeds could be planted and new food would again grow. If there was no rising Sun in the sky after winter, people would starve, civilizations would end. Up until the Industrial Revolution, most people grew their own food out of necessity whereas today, people labor and pay taxes to have just-in-time, mostly synthetically altered food, shipped from halfway around the world.

The luminaires, as the ancients called the celestial beings, were their roadmap to understanding their cosmic relationship with the heavens above. Astrotheology, or the study of the God(s) in the stars, played prominently in the dates the pagans (Spiritual believers pre-Western religion) was set for sacred rituals and sacred rites to honor the role of celestial beings that so graciously gave live to all.

"Paganism is a term that derives from Latin word pagan, which means "nonparticipant, one excluded from a more distinguished, professional group." The term was used in the 4th century, by early Christian community, in reference to populations of the Roman world who worshipped many deities." ~Wikipedia

The pagan legend of the "Sun of God" rising from the cross begins on the Autumn Equinox (equal night and day) or September 21st of each year. From the Northern latitudes in the Fall, the Sun, on its outer movement towards the outer regions of Earth, was observed by the ancients to go down and *hang* on the Southern Cross constellation for three months. The Sun of God was said to hang for three months on the cross until December 22nd, when the Sun then died on the cross for three

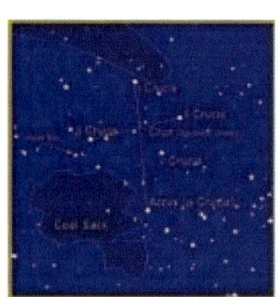

days until it began to once again rise on December 25th.

The Roman Catholic church took over December 25th to celebrate the birthdate of their savior, Jesus Christ the "Son of God" while at the same time burying the pagan celebration of the "Sun of God" celebrated pre-Roman Empire.

The Roman Catholic church, by issuing the Gregorian calendar in the mid-16th century, eliminated the pagan Lunar calendar of 13 months, with 28 days per month, marking one-years' time. They created the Sun calendar we all use to celebrate mass on a Sun-day. This is also why NASA chose the Sun God, Apollo, a phallic missile. as their symbol for the Moon missions. This is also why there are phallic male obelisks in every major western center of governance near an "oval office" of some type.

Once the Romans had conquered pagan sacred temples around Europe, the Church then built temples directly over many of the same pagan sites which are situated directly over energy ley lines on Earth. The New Year calendar date for the pagans was April 1st when crops would begin to grow anew. The Roman Catholics changed the pagan start of the new year to January 1st, after the two-faced God, Janus and renamed the pagan new year day, April Fool's Day.

The word "Easter" comes from the position in Spring of the Eastern star. For pagans, Spring was to mark when seeds could be sowed so that food could then grow and life could go on. East-star Sunday, the day of the resurrection of Jesus Christ celebrated by many western religions, is determined on the first Sunday, after the first full moon, after the first day of Spring. Few western religions will acknowledge the vast Astrotheology in their Christian teachings. The three stars in the belt of the Constellation of Orion are in biblical terms, the three wise kings who followed the brightest star, Sirius, to the birth of Jesus Christ.

The three kings or three magi effectually "follow" the star in the East to the manger, the birthplace of *God*'s Sun at the Winter Solstice. The three gifts of the magi are Frankincense, Myrrh, and Gold. Frankincense is an amber resin that was burned at solar temples, Myrrh was known as "tears of the Sun," and Gold too long represented the Sun in the ancient world.

The recurrent "virgin" theme in Christianity represents the constellation Virgo, which is Latin for virgin. The ancient glyph for Virgo looks like an M which explains the M names of "virgin mothers" like *Jesus*' mother Mary, Adonis' mother Myrra, Buddha's mother Maya, and Horus' mother Isis Meri. Virgo is also called the House of Bread and the zodiacal symbol shows a woman holding a chaff of wheat, representing the August/September time of harvest. Bethlehem also means "House of Bread" and is a reference to the constellation Virgo, not a place on Earth.

Even our astrological signs themselves denote the growing of food. Aries, a Rams horn's, looks like a sprig, or 'spring,' of a new plant rising out of the soil. The next astrological sign is Taurus, the Bull, who plows the fields in May. Next, in June, the twins of Gemini harvest and sow the fields of good and plenty, etc. All astrology, or study of the stars, is Flat Earth based yet you'll find very few astrologers who will even acknowledge this fact…yet. When astrologers read someone's chart they cast the individuals annual Solar Returns. The Sun is calculated to have moved back to its same position as your birth date, not an "Earth Return."

"Since the Earth experienced 4 different seasons, all the same and equal (in time) each year, the round Sun calendar was divided into 4 equal parts. This is also why we have, in the Bible, only 4 Gospels. Of this point, there can be no doubt. The 4 Gospels represent the 4 seasons which collectively tell the entire story of the life of *God*'s Sun. Matthew, Mark, Luke, John are Spring, Summer, Autumn and Winter. This is why the famous painting of 'The Last Supper' pictures the 12 followers of the Sun in four groups (of three) the seasons!"
Jordan Maxwell, (http://www.jordanmaxwell.com/articles/astrotheology/astromain5a.html)

*****

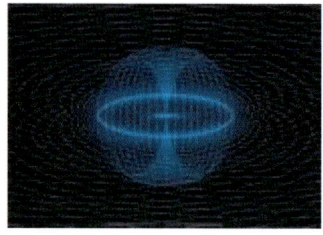

The word "world," or "whirled" comes from the energy vortex created by the electromagnetic battery of our Flat Earth. This plasma field of electrical charges emanate from the North Pole negative charge, or anode with the Antarctic Circle positive charge, or cathode. The salt in the ocean provides the electrolyte catalyst to allow for alternating currents through Earth's plasma battery field. This is what Nikola Tesla proved with his Tesla coil, in his electromagnetic experiments he was able to create artificial lightning storms. He found a way to use Earth's plasma field to generate free energy for all. Sadly, capitalist J.P. Morgan, who was financing Tesla's projects, abruptly stopped financing the project because the banker could not make profit from free energy that Tesla proved was everywhere and accessible to all. Those that run our world will not give us the free energy developed from Tesla way back in the early 1900's. After Tesla's death, the US military confiscated Tesla's papers on free energy while mocking us some 100 years later with the Tesla car run on batteries. Free energy for all would likely end all wars and create world peace in our lifetimes. Most are completely unaware that we had the knowledge and technology, thanks to Mr. Tesla's experiments, to produce free energy for all. Why are we not taught this in our schools or science classes?

Humans are also contained in a plasma toroidal field of perpetual regeneration inside a closed system of a vortex, or whirling energy. Like Earth, humans are self-enclosed batteries as well.

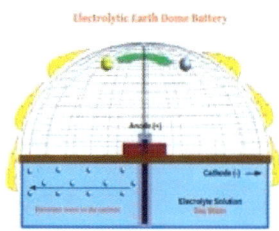

When someone is dying in a hospital they will use electrical stimulation paddles of positive and negative energy to attempt to revive the person. When they arrive at the hospital they are put on intravenous saline solutions to keep the electrical charge. Doctors use paddles to resuscitate someone who is in cardiac arrest for this reason. What do they use to resuscitate the patient? Electric currents; just like the sea that has currents. What else do doctors do? They perform surgery, because it is an electrical surge to the body of the patient, thus surge-ery.

We can see our own battery fields called auras, or light energies, through Kirilian photography. These auras, or known by Christians as halos, are directly connected to the energy chakras in our body. When Kirilian photography is used to show one's aura, often it will show different colors for different people. This represents the different vibrational chakra frequencies that are dominant in each one of us.

*****

For over 100 years most world education systems have taught little-to-no Flat Earth history and cosmology thereby keeping us disconnected and uninformed from our rich historical ancient past. It was not until November of 2014 when an obscure, little known 32-year-old American, named Eric Dubay, published his book, *"Flat Earth Conspiracy"* that the Flat Earth subject began to be purposely introduced to the public at large. Mr. Dubay runs the website, "Atlantean Conspiracy" as well as having a book written by the same title about conspiracy and occulted events in history. Shortly after his book release, at the onset of 2015, the Flat Earth story exploded onto the Internet and became one of the top trending stories that year, according to Google trends. Prior to reading Mr. Dubay's book, I personally had never heard of Flat Earth cosmology but soon researched and educated myself to discover, and uncover, some very convincing evidence of a massive lie as well as a beautiful narrative about our place in the Universe that was known by most ancient cultures, yet hidden from modern history.

"It is easier to fool people than to convince them they have been fooled" ~ Mark Twain

Like many of you, the more I tried to disprove geocentrism, the more convinced I became of its authenticity. This led me to discover and uncover the massive 500-year heliocentric lie. A huge lie told and sold by the Roman Catholic Church to keep most of humanity feeling insignificant and just one of billions upon billions of likely universal inhabitants while also allowing for what many believe to be, a coming faked alien invasion, all perpetrated by the Vatican.

I found it also very interesting and curious when I learned that there are several recordings of US Presidents and Congressmen speaking specifically to Flat Earth as far back as 2011, well before anyone had even heard of the Flat Earth subject matter. There is also Flat Earth references in cartoon shows like 'Family Guy' as well as many Flat Earth maps hidden in plain sight in movies like 'Back into the Future' and movies made about the Jesuits, like "Men in Black" starring 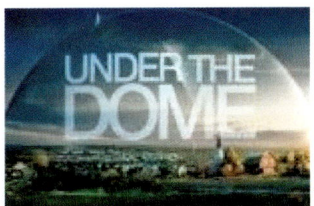 mainstream actors like Will Smith and Tommy Lee Jones, that reference Flat Earth as well. Coincidence or a great Cons-Piracy?

In the Fall of 2016, Paul Michael Bales provided conclusive documentation that the then 32-year young Mr. Dubay had received $100,000 compensation from the National Vaccine Injury Center (NVIC) directly after a two-month long stint in Washington D.C. previous to his release of his Flat Earth Conspiracy book. Mr. Bales claims it was he who provided books for Mr. Dubay's research on Flat Earth. Mr. Dubay professes to be a vegan health nut and against all vaccines. So was this possibly Mr. Dubay's payoff to get the Flat Earth meme into the public arena by his handlers? Was releasing the FE meme another divide and conquer strategy by the powers-that-be like we are seeing today with other government created controversies like LBGT, #Blacklivesmatter, Trump vs. Islam and the Immigrants, etc.? Mr. Dubay also has had a self-publicized romantic partner who is said to have high level connections to generals in the Thai military. He often rips into Freemasons for the perpetuating the heliocentric lie, while "forgetting" to inform all that his Uncle is also a high-ranking Freemason.

As the FE movement grew into 2016, Dubay resorted to cross dressing in his videos and verbally ripping into those that are gay, that eat meat, believe in dinosaurs or nuclear power. He also alleged that Hitler was not actually such a bad guy after all while making mock-u-mentaries about other Flat Earth researchers. All these distractions he self-created despite proclaiming to all that his sole purpose was to get the Flat Earth truth out as bring down the empire…but not the Roman Empire. He refused all calls to link the Vatican with the Flat Earth conspiracy even after so much evidence was presented to him about their involvement.

******

Flat Earth Society Vandalism

From first grade on, nearly every public classroom has had a round ball Earth globe. We have global corporations, global partnerships, and global interests. We span the globe in our wide world of sports and go around the Sun each year to mark our birthdays. Each day, we are all shown perhaps hundreds of round-ball Earth images and logos in media, movies, and advertisements. We have been pre-conditioned to accept these globe ball images as real. These images get tattooed onto our subconscious minds, reinforced over and over again by social media.

Few are even aware that the US government, and other governments around the world, spend billions of dollars each and every year on what they call, "brain mapping." Brain mapping studies very scientifically what goes on behind our eyes and between our ears. Business calls it marketing, getting someone to buy your product. Those in power use brain mapping for a much more devious and secretive purpose, mind control of the masses.

Every year, those that wish to control all hire the best and brightest PHD graduates to assist in their research on human behavior. They use the latest algorithmic super, quantum, and Dwave computers to better understand how to stimulate and manipulate the human brain for desired outcomes. What do you think those in power would be able to program into us all if all relied on social devices for what they know about the world, reinforced by a lock, stock and owned public education system, that begins in "pre"-school? Nearly all media is owned by the wealthy elite. They give us daily memes for community discussions and school discourses with the right hand, while the left hand furthers their self-proclaimed desire to control all and eliminate those that no longer server their agendas through silent weapons for quiet wars.

"In our dream we have limitless resources, and the people yield themselves with perfect docility to our molding hand. The present educational conventions fade from our minds; and, unhampered by tradition, we work our own good will upon a grateful and responsive rural folk. We shall not try to make these people or any of their children into philosophers or men of learning or of science. We are not to raise up among them authors, orators, poets, or men of letters. We shall not search for embryo great artists, painters, musicians. Nor will we cherish even the humbler ambition to raise up from among them lawyers, doctors, preachers, statesmen, of whom we now have ample supply."
  - Rev. Frederick T. Gates, Business Advisor to John D. Rockefeller Sr., who founded the US General Education Board in 1903

The world education system was set up by the wealthy-elite at the turn of the Industrial (D)evolution. The three richest men of the Industrial Revolution; Astor, Carnegie and Rockefeller held 30% more wealth than the richest men today of Gates, Buffet and Ellison. With their massive wealth, they bought influence wherever it was needed. In government, the media and to create a public education system to produce obedient workers. Those in power only wish to have good obedient tax paying workers. Never would they wish to train others to take away their power base, so the illusion of capitalism was created where raping the land, bankrupting a neighbor or deceiving a world population with lies, was "just business."

They instituted the progressive tax system in 1913, the same year the for-profit private corporation, Federal Reserve, was created. This kept those making more money from ever coming close to mega wealth enjoyed by the Robber Barron's still in power today. To avoid taxes and to influence government and education, they exploited the non-profit organizations, or NGO's. The Rockefeller, Guggenheim, Morgan and Bill and Melinda Gates Foundations among others, still funds and still hold large sway to this day over public policies and what is educated to the children.

"Once in power, we stay in power" is one of elite's mottos. Another is. "Prey or Prey," and "You are either at the table, or on the table." The effect change through "Order out of Chaos," such as the directly immediately after the events of 9/11/01 when the Patriot Act in the United States was enacted which subverted basic civil rights.

Most who cannot critically think for themselves are sold what to think and how to behave. This is why we buy "smart" phones and install "smart" meters on our homes. These radiation devices monitor everything we do and say while emitting dangerous radiation that increases the risk of cancer, which benefits Big Medicine. Meanwhile, they use our own desires for extreme convenience to enter our homes to eavesdrop and record our most private conversations with devices like Amazon's Echo, Google's Home, Microsoft's Cortana and Apple's talking head, Siri. Please understand, dear reader, technology is extremely more advanced in all areas of mind control than most are aware. The internet was released by the advanced military arm of the US government, DARPA in the mid-1990's for a reason and we have had found ways to harness free energy from Earth's electromagnetic field for over a century while still using hundred-year old combustion engine technology in cars.

*****

"Signs and symbols rule the world, not words nor laws." ~ Confucius

Our minds record what our eyes see, even if we are not consciously aware this is happening. Even the best and the brightest are said to only be using up to 15% of their brain's capacity. The subconscious mind can be programmed very easily through symbols that can escape the conscious mind. This is called subliminal programming and is used very effectively by many advertisers to suggest into your subconscious mind that you want or desire their product they are selling.

Colors are also used to plant memes in the mind. Blue being calm, purple being royal and red being anger. Our mind records the image that we are seeing but our awake mind may not see it. Can you see the double cross in the Exxon logo? It's all about secretly implanting thoughts and memes into our heads, though most are not even aware that much of their actions and understandings have been subliminally implanted into our brains over social media devices and the like. Also,  using certain words in a rhythmic pattern can alter thought processes. In scientific terms, it is called neuro-linguistic programming or NLP. The bottom line is that the powers who control our government, also control what is taught in schools and what is broadcast as news. They can, and do, control what most think, say and act. To speak against such deeply seated programming, like the Flat Earth subject, immediately breeds contempt and scorn for most who don't even understand the most basic of astronomy, physics of empirical evidence gathering.

"Why do we pay taxes?"

A simplest example to show how we are all mind controlled is to ask anyone why they pay taxes to their government even though the wealth-elite and many, many corporations pay little to no taxes whatsoever while making huge profits. We pay our labor in taxes, we pay taxes for our consumption and get taxed when we die. We pay taxes to drive our car, for gas for our car and to build the roads for our cars. We pay sales taxes, property taxes, and even sin taxes (cigarettes, alcohol, etc.). Most work the first 4 ½ months of each year just to pay their income taxes from labor they provided! All the while our "leaders" are receiving single digit approval ratings from the same tax payers. They spend trillions of dollars to put their own people in massive debt, run up the stock market for more wealth to the wealthy, and invade mostly brown colored people's countries under false pretenses of WMD's… killing millions. The answer we all give is the exact same. "I pay my taxes so I won't go to jail!" Barnum and Bailey Circus would be proud of the conditioned responses the mind control specialists have enacted on the many by the very wealth few. Taxation is a form of slavery when enforcement is needed to get people to pay their taxes, yet due to subliminal messaging we all pay and pay for our own enslavements.

Many corporate logos, as well as the 'images' we are shown from space, show a blue-marble Earth. As recently as 2015, NASA has had to admit that the images of Earth are just that, images. Not a single one frame photograph of Earth has ever been produced by NASA. Not one over the decades of alleged space travel. Flat Earth activists have conclusively shown how many Earth images provided by NASA have been computer generated images (CGI) or photoshopped. Rob Simmons, an employee of NASA, said in a statement in 2015, the following: "My part was integrating the surface, clouds, and oceans to match people's expectations of how Earth looks from space. That ball became the famous Blue Marble."

Why are there no live streaming video cameras on the Moon showing a live picture of an entire Earth revolving at 1,000 mph? NASA can allegedly transmit full photographs from Mars and even Pluto, from millions of miles away using tiny microwave radio beams, yet we do not have one live camera of full Earth or even one unaltered photograph?

The most likely reason NASA has not conducted, or concocted, another mission to the Moon is because camera technology got too advanced for all the tomfoolery they were able to pull off in the late 1960's and 70's during the faked Apollo Moon missions.

The images of space travel to Moon was implanted back when NASA was created in 1958 and secret society member, Walt Disney, started putting out Men to the Moon movies. If you Google "satellites in space" you will also only find CGI images, no real photographs of the ISS space station or the other thousands of satellites said to be up in space, only images. Why? On Google Earth, there are no real photographs of the Arctic or the Antarctic, either. Why, if we have such advanced technology we can send a picture from Pluto, millions of miles away with just a wireless radio beam?

NASA announced in 2006 that it accidentally erased all Apollo moon footage, over 200,000 tapes! They claimed in all seriousness that the tapes got erased, or degaussed, because they needed the magnetic tapes for other recordings! They then announced in 2009 that they had been able to digitally restore just a few of the Apollo moon footage. With an unquestioning and unaware public, and no accountability to anyone, NASA can just do and say whatever they wish, no matter how absurd the story being told/sold is.

The International Space Station (ISS) is composed of a fragile, solar-paneled spacecraft said to be orbiting Earth at incredible speeds of 17,500 mph in geosynchronous orbit some 210 miles above Earth in the Thermosphere. The Thermosphere is where temperatures are said to reach upward of 2,500°F or 1,300°C! We are sold on believing that solar panels, space walks, etc., can occur in such unearthly and inhuman thermal temperatures much less go outside for a spacewalk to install solar panels or whatnot.

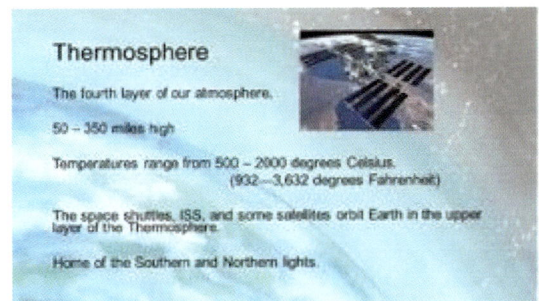

It is also worth wondering how the astronauts and spacecraft can avoid the hundreds of thousands of space debris and graveyard satellites in similar orbit and traveling at speeds of over 6,000 mph. NASA acknowledges that if an astronaut was hit by even a small paint fleck of a space debris, death would likely result. Even a hit to the spacecraft window at 6,000 mph on the space shuttle, or space station, would also cause instant death to all aboard. There have been zero reports of anyone ever haven gotten even nicked by hurtling space debris, even after decades of space walks, and space travel, involving hundreds of missions into space.

The Law of Ridiculous Numbers

We have all been schooled, like fish, to believe that we orbit once around the Sun every year only to return to the same nearly exact position for our birthdays. Few can recite even the most basic of the physics alleged to be behind the incredible spin, rotation, wobbles, elliptical orbits, and movements throughout the heavens that heliocentric theory proposes is occurring throughout our universe. Those that wish to have us believe heliocentrism have simply made the numbers over time, so huge, so enormous, that humans have no relative experience to comprehend what such large numbers even mean. NASA can make up any number they want, and no one can question the validity of their outrageous statements of "facts".

According to modern astronomy, and unchallenged for over a century, is that the Earth is spinning on its axis at over 1,000 mph at the Equator. Earth is said to cover some 24,000 miles at our beltline in 24 hours (24,000/24 hrs. = 1,000 mph). However, we never feel even a whisper of a wind of such speed and movement. How can that be? No one can even tell you what makes the Earth spin on its axis at such speeds in perfect uniformity and precision that Earth's spin can be timed down to the nanosecond. Few can tell you what even made the Earth spin in the first place or why Earth settled at 93 million miles away from the Sun to create the perfect weather for human habitation.

To journey around the Sun each year we necessarily must be traveling through space over 533 million miles. This means our Earth, while spinning at 1,000 mph on its axis, must also be moving at over 1,000 miles per second to go around the Sun! Most humans can only physically relate to speeds of 100 mph or so in a car. Stick your head out the car window and you can hardly breathe.

Now imagine speeds 10X faster that the Earth is spinning and 100X faster that the Earth is trucking just to get back to our birthdays each year. Again, the numbers are beyond what humans can relate to, so trust is mandatory in the science community to their magnitudes of speed or Earth is moving.

Add to these incredible speeds of Earth travel of our Solar Systems movement around the Milky Way Galaxy. This would mean that all in our Solar System are also hurtling through space at speeds of over 500,000 mph. Yet the stars above remain fixed, no matter how many centuries of travel the Solar System is said to move. How can this be?

The Hubble telescope is said to send out radio frequency (RF) waves that travel vast distances to record a planet's size, velocity, rotational speed as well as its make-up of water, gases, and orbital relationship to other stars. NASA says they learn all this data from a tiny radio beam shot out into far space. Some of these distances are said to be up to a quadrillion miles, out and back. Again, numbers none of us can relate to. The Hubble telescope is credited with finding thousands of exoplanets, or Earth like planets. This is an amazing feat by the tiny micro radio waves that must go through space dust, comet tails, time and space warp gravitons, and the mass rocks in the asteroid belts, without so much as a deviation, or a wave interference, to accurately capture and record such precise measurements.

The current physical distance we are told of our galaxies width and breadth, measured solely by infrared microwave and radio frequency beams, is just over 46 billion light years, according to NASA. If we take their numbers and measure in terms we can be familiar with, we get a number of 276,000,000,000,000,000,000,000 miles away! Just trust the science we are told, yet we can lose a radio signal in our car while just driving around.

The second Law of Thermodynamics, also known as the Law of Entropy, along with the principles of resistance and drag, prove the impossibility of Earth being a uniform spinning ball. Over time, the 1,000-mph spinning-ball Earth would experience significant friction and drag constantly slowing Earth's spin. However, only the very slightest change of speed of Earth's rotation has ever been recorded over as long as science has measured Earth's spin. The rotation of Earth is said to be so uniform we can time its rotation to a nanosecond. The Earth's rotation is explained as "angular momentum," where perfect speeds are kept without slowdown over centuries. Science explains to us that there is no change of Earth rotational speed due to there being no friction or drag over millions of years' time. They thusly discount the movement of the oceans, the anthropogenic influences of man and machine, the flowing of the great rivers and the changes in our atmosphere. Nothing effects the spin on the spinning ball, we're told.

"Every experiment ever designed to detect the motion of the Earth has failed to detect Earth's motion and/or distinguish it from relative counter motion of the universe."
—*Mark Wyatt*

Why do we never feel so much as a wisp of such massive motions of Earth through space and time? Clouds and butterflies meander this way and that way, while chimney smoke lofts straight up, never being affected by Earth's spin it seems. Gravity holds all in on the spinning ball, including the mighty oceans, yet when we wish to defy gravity and jump up in the air, we can at will on our magical spinning ball Earth, we're told.

The indoctrination to the masses of a heliocentric story was interrupted at the turn of the 20th century when experiments, such as the Morey and Michelson experiments in 1887, proved conclusively the existence of Aether, thereby proving the Earth was still and stationary. To save the heliocentric theory from being totally discredited by the experiments performed by Michelson, Morley, Gale, Sagnac, Kantor, and others, establishment master-minds helped Albert Einstein create his Special Theory of Relativity. Mr. Einstein simply created a fifth element of nature called 'time' so that the heliocentric theory hoax could continue. He successfully banished the absolute Aether/firmament from scientific study in science and school learning. He replaced the Aether with a form of relativism to time which allowed for helio-centricism and geocentricism to hold equal merit. Forget, he was telling us, that time is a man-made construct, not an element of Nature.

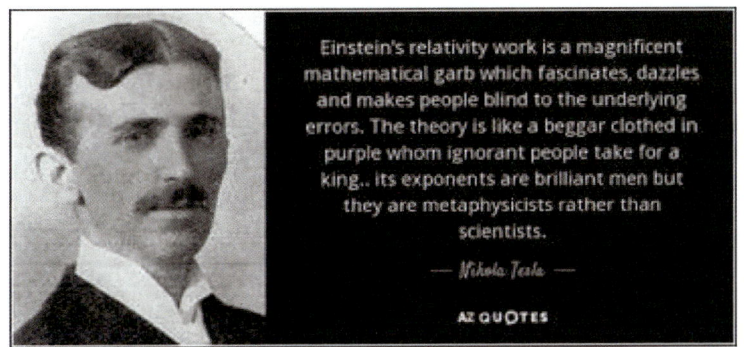

If there is no universal Aetheric medium where all things exist, then philosophically one can postulate complete relativism about the movement of two objects such as the Earth and Sun. Nowadays, just like the theory of helio-centricism, Einstein's Theory of Relativity is accepted worldwide as gospel truth, though Mr. Einstein himself admitted geocentricism is equally justifiable.

The entomological meaning of the word planet is 'wandering star.' Our Plane-t is a plane, which is why to this day we still call the oceans, Sea-*level,* and flying crafts are called aero-*planes* and Earth's mantles are called *plate*-tectonics. Ancient Greek etymology uses the letter "T" for Terra, or land, so a Plane-T means a plane land or Flat Earth!

Admiralty Law Words

Examples well quoted are:

THE HOLY SEE (SEA)

Birth Certificate - Berth Certificate	Court: Dock - Harbour: Dock	Your Worship
citizenSHIP (14ᵗʰ Amendment citizen)	Survivorship	Floating your company
LeaderSHIP	Cash Flow	Deposit
Withdraw	Liquidisation	Bank
Currency	Floating	Loan Sharks
Liquid Assets	I have my head above water	Frozen assets
Your account has been frozen	apprenticeSHIP	Bank Balance
Bail	Bail bond	Bailiff
Bankruptcy	Bankruptcy law	Bar
Bar examination	Disbarment	Discharge
Seal	Sidebar	Ownership (name in all CAPS)

- ambassadorship - authorship - battleship - censorship - championship - companionship - craftsmanship
- draftsmanship - fellowship - friendship - hardship - horsemanship - internship - membership
- partnership - relationship, - scholarship - sponsorship - sportsmanship - warship - workmanship

History, it is said, is written by the victors. Since the Vatican has been in power of world events since the days of Caesar, they have controlled the collective narrative of the very language we use and the words of law that govern us. (See Appendix III on Deceptive Language)

They use legalese word trickery to create legal fictions of souls at Sea, or strawmen. We go to court on our citizen-*ship* where a *bail*-iff lets us through the gated, *docket*. We go take our deposit *slips* to the *river* bank where we account for our holdings referred to as *liquid, illiquid* and *frozen* assets… and so on it goes. We are all considered, "chattel" and "lost souls at Sea" claimed by the Vatican as under their trust and jurisdiction as Trustor of Jesus's holdings, since he is not here to administer his owner-*ship* of lands and souls.

The Roman Latin Laws set forth humanities enslavement and owner of all lands on Earth by the Roman Catholic Church was created with the very first Cest Qui Vie Trusts beginning with Pope Bonaface the VIII, in 1302.

In that year, Pope Boniface VIII (1294-1303) issued his infamous Papal Bull *Unam Sanctam*, being the first Express Trust claiming control over the whole planet and effectively "King of the world." In celebration, he commissioned a gold-plated headdress in the shape of a pinecone, with an elaborate crown at its base. The pinecone is an ancient symbol of fertility and one traditionally associated with Ba'al as well as the Cult of Cybele. The three tiers represent the three trusts that claim rights of ownership over all.

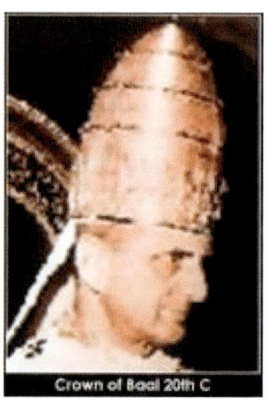

Crown of Baal 20th C

The next three Papal Bulls, or Canon Law trusts created by the Vatican, were administrative in nature and included the term "in perpetual remembrance" which means - forever.

1st Crown of Crown Land

The first Testamentary Trust was the "Roman Panifex" in 1455. This Bull had the effect of conveying the right of use of the land as Real Property from the Express Trust Unam Sanctam to the control of the Pontiff and his successors in perpetuity. Hence, all land is claimed as "crown land."

2nd Crown of the Eternal Crown. The second Crown was created in 1481 with the papal bull Aeterni Regis meaning "Eternal Crown" by Sixtus IV. This 2nd Crown is represented by the 2nd cestui Que Vie Trust created when a child is born. This being the sale of the birth certificate as a Bond to the private central bank of the nation, depriving them of ownership of their flesh and condemning them to perpetual servitude as a Roman person, or slave.

3rd Crown of the Ecclesiastical See The third Crown was created in 1537 by Paul III through the papal bull Convocation, being the third and final testamentary deed and will of a testamentary trust, being the trust set up for the claiming of all "lost souls," or lost to the See.

The Venetians assisted in the creation of the 1st cestui Que Vie Act of 1540 to use this papal bull as the basis of Ecclesiastical authority of Henry VIII, King of England. This Crown was secretly granted to England in the collection and "reaping" of lost souls. The Crown was lost in 1815 due to the deliberate bankruptcy of England and granted to the Temple Bar, which became known as the Crown Bar, or simply the Crown.

The Bar Associations have been responsible ever since in administering the "reaping" of the souls of the lost and damned, including the registration and collection of Baptismal certificates representing the souls collected by the Vatican and stored in its vaults. This 3rd Crown is represented by the 3rd cestui Que Vie Trust created when a child is baptized being the grant of the Baptismal certificate by the parents to the church or Registrar being the gift of title of the soul. Thus, without legal title over one's own soul, a man or woman may be "legally" denied right to stand as a person, but may be treated as a creature and thing without legally possessing a soul. Hence, why the Bar Association is able to legally enforce Maritime law against men and women- because they can be treated as things, cargo, or chattel, that does not possess a soul.

*****

~ from the Independent.co.uk, 10/28/14

"The theory of evolution and the big bang are real, and God is not 'a magician with a wand', Pope Francis declared at the Pontifical Academy of Sciences. He made comments which experts claim put an end to "pseudo theories" of creationism and intelligent design, some of which were encouraged by his predecessor, Benedict XVI, as reported by the *Independent*.
The Pope explained that the scientific theories weren't incompatible with the existence of a creator, rather, they 'require it'. When we read about Creation in Genesis, we run the risk of imagining God was a magician, with a magic wand able to do everything. But that is not so. The big bang, which today we hold to be the origin of the world, does not contradict the intervention of the divine creator but, rather, requires it.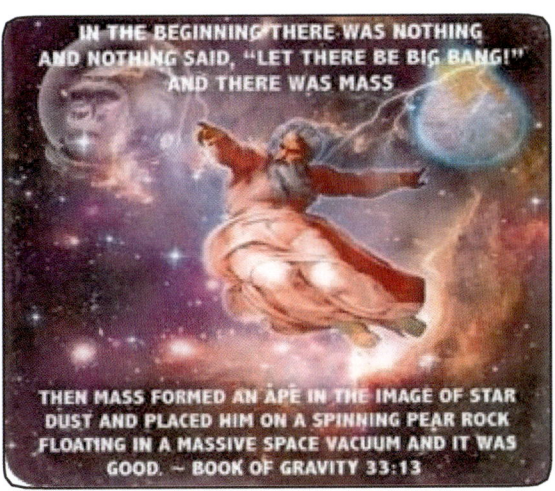
Evolution in nature is not inconsistent with the notion of creation, because evolution requires the creation of beings that evolve.

The Catholic Church has had a reputation for being anti-science. However, Pope Francis' comments were in line with the progressive work of Pope Pius XII, who actively welcomed the big bang theory and opened a conversation on evolution. In 1996, John Paul II went further and suggested evolution was 'more than a hypothesis' and '**effectively proven fact**'.

Yet more recently, Benedict XVI and his close advisors apparently endorsed "intelligent design" underpinning evolution – the idea that natural selection by itself is insufficient to explain the world's complexity. Giovanni Bignami, president of Italy's National Institute for Astrophysics, told the Italian news agency Adnkronos, "The pope's statement is significant. We are the direct descendants from the big bang that created the universe. **Evolution came from creation**." So now the circle has been completed over 500 years. Church and Science are in harmony. We came from nothing that created everything proven by science and confirmed by the closest authority to God, the Roman Catholic Church. Got it!

Intelligent Design?

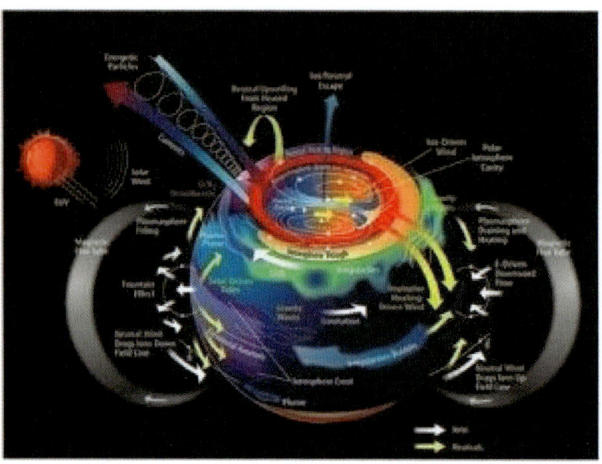

The National Aeronautical Space Administration (NASA) was created in 1958. They, along with the US Naval Intelligence and the Vatican World Assistancy, have controlled all news from space since that time. All other space agencies, in China, Japan, India, and Russia, etc., are all are subsidiaries of NASA. The very first rocket builder for NASA was a high ranking German Nazi named Werhner Magnus Maximilian Freiherr Von Braun. Mr. Von Braun was the 'wunderkind' chief designer of the V1 and V2 rockets that rained hell fire on Europe during WWII, killing thousands of allies. He and many other Nazi scientists, many of whom who practiced eugenics and torture tests on live humans, were brought over to the US with full assistance and knowledge of the US military and high-ranking government officials over decades of time. In all, from the 1950's to the 1970's, under "Operation Paperclip" (Paper clips were used to identify those being transferred), over three thousand high-level Nazi scientists were brought into the US after World War II. The Nazi scientists were issued Vatican passports, then shipped through Vatican ratlines to South America, then brought into the US.

These men were some of the most heinous criminals of WWII who then went on to run top US corporations like Bayer Drug Corporation, (who recently purchased Monsanto, creating the Farmacology company from hell). Many of these scientists were used to conduct human experimentation through projects like MK Ultra. (MK, the German spelling for Mind Kontrol). Mr. Von Braun ran the Saturn rocket program all the way through the alleged Apollo moon missions. Interestingly, Mr. Braun's gravestone cites Psalms 19:1 from the King James Bible referring to the 'firmament' above. Were his last words he left on his gravestone to tell us we were really living under a dome?

The first administrator of NASA in 1958 was Thomas Keith Glennan. His previous experience was at Paramount Studios as a movie producing executive. Before that, he helped run the US Navy's Underwater Sound Laboratories during WWII in New London, Connecticut. One could easily conjecture that this was the perfect training of a NASA Administrator to make a movie about fake space missions using underwater studios to film for weightlessness effects in space with a green screen background. NASA had Photoshop and access to secret military technology long before it was brought to the public.

In 1966, Star Trek came to TV from the same Paramount studios that Mr. Glennan had worked as movie executive. The show's creator, Gene Rodenberry, was a 33rd degree Freemason. The show introduced flat screen TV's, tasers that stun, medical tricorders, cell phones and talking computers. None of these technologies was then available to the public but are every day devices since the turn of the 21st century. This will give you an idea of how long those in control have had the technology.

Walt Disney and his studios were also very heavily involved in the early years to help pull off man to the moon hoaxes, one of the greatest mind control deceptions and hoaxes in over 500 years. While Von Braun was starting up the Saturn rocket program at NASA in 1958, he was also being used by Walt Disney (Priory of Scion) to begin the selling and promotion of space exploration through TV programming. Mr. C. Fred Kleinknecht was head and Chief Administrator at NASA at the time of the Apollo Space Program. After he completed running the first Apollo missions, he resigned and was immediately named the Sovereign Grand Commander of the Council of the 33rd Degree of the Ancient and Accepted Scottish Rite of Freemasonry of the Southern Jurisdiction, located in the heart of Washington D.C. Apparently, this was his reward for pulling off the moon mission hoaxes that were created in studios like Lookout Mountain above Laurel Canyon in Southern California. Many surmise that movie director, Stanley Kubrick, creator of the film 2001 Space Odyssey, was used to help authenticate the fake moon missions and then was later killed to keep the big secret from coming known to the public

Many astro-nots were either high-ranking members of secret societies or came from families of Freemason lineage such as: Buzz Aldrin, Jr., the second man to lie about walking on the moon is an admitted, ring-wearing, hand-sign flashing 33rd degree Mason from Montclair Lodge No. 144 in New Jersey. Edgar Mitchell, another supposed moon-walker aboard Apollo 14 is an Order of Demolay Mason at Artesta Lodge No. 29 in New Mexico. James Irwin of Apollo 15, the last man to lie about walking on the moon, was a Tejon Lodge No. 104 member in Colorado Springs. Donn Eisele on Apollo 7 was a member of the Luther B. Turner Lodge No. 732 in Ohio. Gordon Cooper aboard Mercury 9 and Gemini 5 was a Master Mason in Carbondale Lodge No. 82 in Colorado. Virgil Grissom on Apollo 1 and 15, Mercury 5 and Gemini 3 was a Master Mason from Mitchell Lodge No. 228 in Indiana. Walter Schirra, Jr. on Apollo 7, Sigma 7, Gemini 6, and Mercury 8 was a 33rd degree Mason at Canaveral Lodge No. 339 in Florida. Thomas Stafford on Apollo 10 and 18, Gemini 7 and 9 is a Mason at Western Star Lodge No. 138 in Oklahoma. Paul Weitz on Skylab 2 and Challenger is from Lawrence Lodge No. 708 in Pennsylvania.

NASA astronauts Neil Armstrong, Allen Sheppard, William Pogue, Vance Brand, and Anthony England all had fathers who were Freemasons as well. This follows in line with the founding of Washington D.C. where Freemason George Washington, in his full Freemason regalia, laid the cornerstone to the United States Capitol building. As you will see by some of the documents provided in this book, NASA is primarily a military operation.

Beginning in 1540, and continuing for over two centuries later, Jesuit priests flocked throughout Near and Far East Asia to teach a heliocentric version of modern scientific astronomy. The real purpose of sending out Vatican astronomers to the "four corners of the earth" was to control the narrative about who we are, why we are here, and where we all came from. In the early 1600s, the Jesuits brought the first telescopes to the United States just as the Copernican theory of a Sun centered cosmology was being "proven" in its initial stages of scientific inquiry. The Roman Catholic, Christopher Clavius (1537–1612), was a high-level astronomer in the sixteenth century. As a Jesuit, he incorporated astronomy into the Jesuit curriculum and was the principal scholar behind the creation of the Gregorian calendar. An alleged scientific calendar that is so inaccurate that a day must be added every four years, then one year skipped. This is why we say there is 365 ¼ days in a year.

Parts of the Roman Gregorian calendar are based upon Roman Emperor Gods like Augustus Caesar named for the month of August and July for Julius Caesar as well as the Roman God, Janus for January.

We have been for a very long time, and still are under Roman Empire rule and legal jurisdiction to this day.

The Roman Catholic Church created our Latin laws used in Common Law, or Admiralty law, in our court systems.

Our money systems are derived from Roman Emperor Julius Caesars first use of coins.

When we fall in love we become *Roman*tic in our *Roman*ce.

All Roads still lead to Rome.

As you will find further in this book, the Vatican priests are Sun occult worshipers. Apollo is their Sun god, also known as Thoth and Osiris, in ancient Egypt and Babylonian legends. The Roman Catholic church practices its Sun occult worshiping in the house of Baal, which is also why they promote the spinning "ball" theory. Even to this day, the Vatican has collaborated with major universities and governments in many countries to jointly build and run telescope observatories to study the heavens. They have even built observatories as recently as 1989, on Mt. Graham, Arizona, called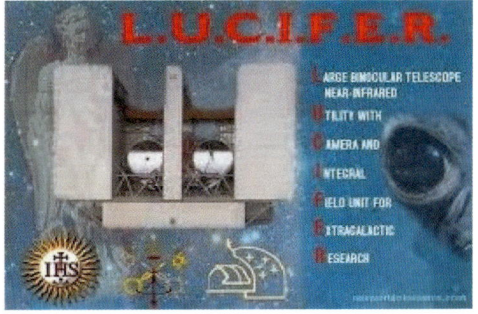
L.U.C.I.F.E.R., which stands for Large Binocular Telescope Near-infrared Utility with Camera and Integral Field Unit for Extragalactic Research.

Nicolaus Copernicus waited until on his death-bed before publishing his work on heliocentric theory. His book was titled, "*On the Revolutions of the Celestial Spheres*." This book, modern historians and astronomers tell us, is said to have changed everything we knew about our relation to the universe. His book was dedicated to Pope Paul III, sanctifier to the creation of the Society of Jesus, also known as the Jesuits. Copernicus became a priest in his later years during the reign of Pope Clement VII (1523-1534). Pope Clement VIII reacted so favorably to Copernicus' newly published book that he rewarded the mathematician turned priest's heirs with a rare manuscript from the immense Vatican archives.

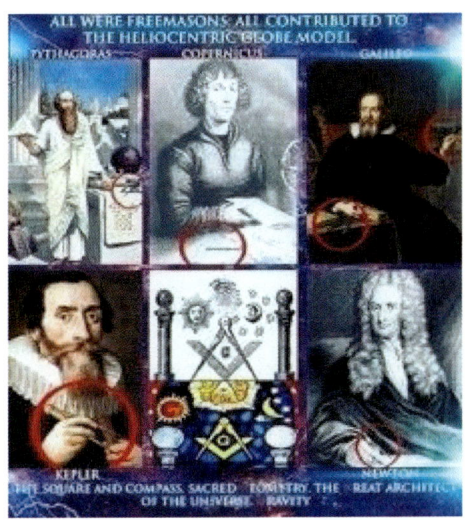

Sir Isaac Newton, once Mint Master for the Queen of England and President of the Royal Institute of Astronomy, is recognized by all of academia and modern science today as the most influential astronomer of all time and a key figure in scientific revolution of the Renaissance Era. "Old apple head," was also a high-ranking member of the secret society of the Priory of Scion, the same secret boy's club that Walt Disney was later a member of. He achieved the high-level rank of Grand Master from 1691–1727.

Newton's book, *Principia*, formed our current understanding of scientific laws of motion and the still unproven theories of "universal gravitation." His work has dominated scientists' view of our physical universe for the last three centuries without question of its validity…. until now. Strangely, Mr. Newton, in all his writings, fails to even mention the word 'weight' in any of his still unproven theories of gravity. The best and brightest of astrophysicists, like Neil de Grasse Tyson, spokesperson for NASA, admits on the record that "we really don't know what gravity is." Nowhere, in any of the Newtonian Laws of Universal Gravitation, will you find the word "weight" to describe gravity, even though the root of the word "gravity" or "gravis", means "weight."

*****

Columbus Day, as we know it in the United States, was invented by the Knights of Columbus, a Catholic fraternal service organization. Back in the 1930s, they were looking for a Catholic hero as a role-model their kids could look up to, the story goes. In 1934, as a result of lobbying by the Knights of Columbus, Congress and President and Freemason, Franklin Roosevelt, who oversaw the occult symbolism on the dollar bill in 1933, signed Columbus Day into law as a federal holiday to honor a Roman Catholic Italian who was one of the worst sailors in all of history!

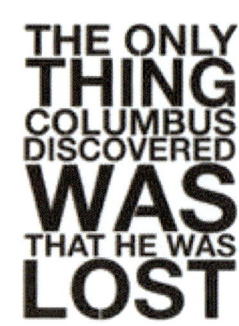

Columbus himself never even set eyes upon the United States on his voyage. The closest he got to the mainland of North America was Puerto Rica. In the aftermath of Columbus's voyage, John Cabot sailed from Bristol, England; which in turn opened the way for the first colony in Jamestown, Virginia and allowed the English to claim America as their own. Columbus was no hero. He can easily be rebranded "The First American Terrorist." When he set foot on that sandy beach in the Bahamas on October 12, 1492, Columbus discovered that the islands were inhabited by friendly, peaceful people called the Lucayans, Taínos and Arawaks. Writing in his diary, Columbus said they were a handsome, smart and kind people. He noted that the gentle Arawaks were remarkable for their hospitality. "They offered to share with anyone and when you ask for something, they never say no," he said. The Arawaks had no weapons; their society had neither criminals, prisons nor prisoners. They were so kind-hearted that Columbus noted in his diary that on the day the Santa Maria was shipwrecked, the Arawaks labored for hours to save his crew and cargo. The native people were so honest that not one thing was missing.

Columbus was so impressed with the hard work of these gentle islanders, that he immediately seized their land for Spain and enslaved them to work in his brutal gold mines. Within only two years, 125,000 (half of the population) of the original natives on the island were dead. Shockingly, Columbus supervised the selling of native girls into sexual slavery. Young girls of the ages 9 to 10 were the most desired by his men. In 1500, Columbus casually wrote about it in his log. He said: "A hundred castellanoes are as easily obtained for a woman as for a farm, and it is very general and there are plenty of dealers who go about looking for girls; those from nine to ten are now in demand."

He forced these peaceful natives to work in his gold mines until they died of exhaustion. If an "Indian" worker did not deliver his full quota of gold dust by Columbus's deadline, soldiers would cut off the man's hands and tie them around his neck to send a message. Slavery was so intolerable for these sweet, gentle island people that at one point, 100 of them committed mass suicide. Catholic law forbade the enslavement of Christians, but Columbus solved this problem, he simply refused to baptize the native people of Hispaniola.

On his second trip to the New World, Columbus brought cannons and attack dogs. If a native resisted slavery, he would cut off a nose or an ear. If slaves tried to escape, Columbus had them burned alive. Other times, he sent attack dogs to hunt them down, and the dogs would tear off the arms and legs of the screaming natives while they were still alive. If the Spaniards ran short of meat to feed the dogs, *Arawak babies were killed for dog food*.

Columbus proves the Earth is not Flat, Seriously?

Columbus's acts of cruelty were so unspeakable and so legendary – even in his own day – that Governor Francisco De Bobadilla arrested Columbus and his two brothers, slapped them into chains, and shipped them off to Spain to answer for their crimes against the Arawaks. But the King and Queen of Spain, their treasury filling up with gold, pardoned Columbus and let him go free.

The real annihilations did not start until the beginning of Columbus's second voyage to the Americas in 1493. For while he had expressed admiration for the overall generosity of Indigenous People and considered the Tainos to be "very handsome, gentle, and friendly." He interpreted all these positive traits as signs of weakness and vulnerability, saying "if devout religious persons knew the Indian Language well, all these people would soon become Christians." As a consequence, he kidnapped some of the Tainos and took them back to Spain.

On his third voyage, in December 1494, Columbus captured 1,500 Tainos on the island of Hispaniola and herded them to Isabela, where 550 of "the best males and females" were forced aboard ships bound for the slave markets of Seville. Under Columbus's leadership, the Spanish attacked the Taino, sparing neither men, women nor children. Warfare, forced labor, starvation and disease reduced Hispaniola's Taino population (estimated at one million to two million in 1492) **to extinction within 30 years**.

Furthermore, Columbus wrote a letter to the Spanish governor of the island, Hispaniola. Columbus asked the governor to cut off the ears and the noses of any of the slaves who resisted being subjugated to slavery. It is estimated that ***100 million Indians from the Caribbean, Central, South, and North America perished at the hands of the European invaders***. Much of that wholesale destruction was sanctioned and carried out by the Roman Catholic Church and various Protestant denominations in willful agreement with the murder, rape and enslavement of the Native peoples of America, truth be told, not sold.

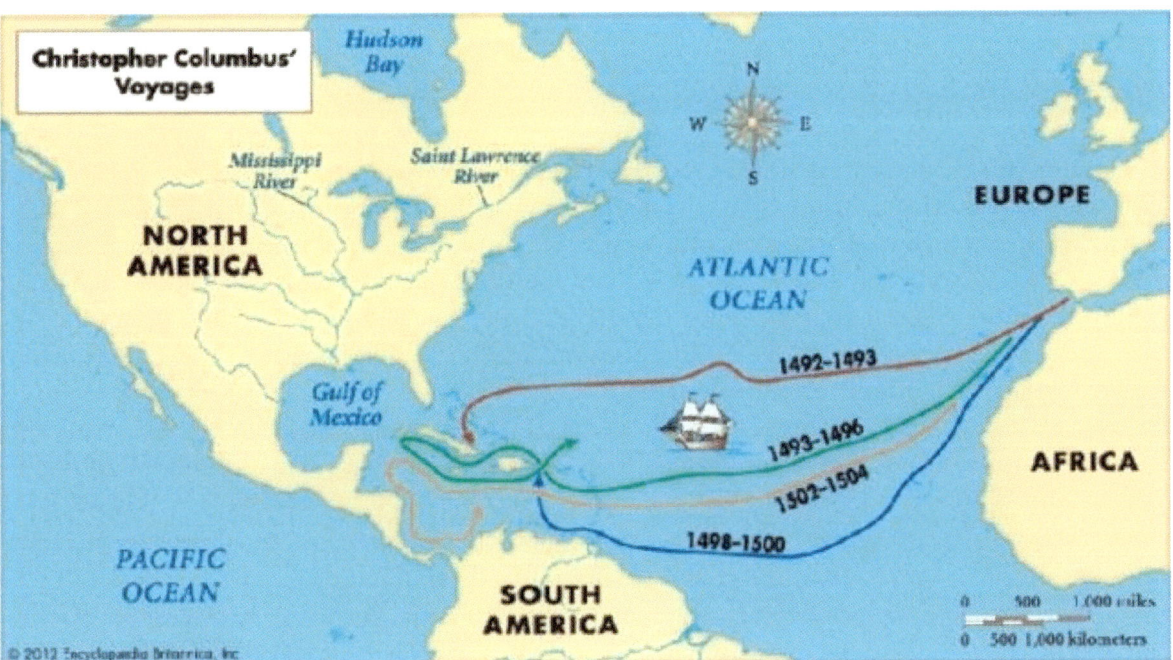

The United States is not America, though those that live in the US call themselves "Americans." America is a continent which includes Canada, Central and South America. To claim the US as America is another psyop to obfuscate and confuse.

When school children learn that Columbus came to America, they think they came to the US, which is a lie on one hand yet a truth on the other.

Minds are like parachutes
They work best when they are open

It is incumbent on all Flat Earth researchers to gratefully acknowledge the heavy lifting and pioneering work of not only the ancient Flat Earth cosmologists but of those who spoke out directly against the Copernicus Theory of Cosmology when the church and state first published their round ball heliocentric theories to all in the mid-late 1800s. These Flat Earth Hall of Fame members include Samuel Rowbotham, Lady Blount, Samuel Shenton, William Carpenter, Gabrielle Henriete, David Wardlaw Scott, Gerard Dickson, Wilbur Glenn Voliva.

Their work was picked up in the 1970s with the fine work of Charles and Marjory Johnson, "the last iconoclasts", who tirelessly published and promoted their newsletter, the "Flat Earth News." As their publication subscriptions began to increase, suddenly their home and all their Flat Earth records were lost when their house was burned down by unnamed arsonists. They died poor and desolate in the mountains outside of Los Angeles, California.

Another Flat Earth hero was Mr. Gerard Dickson who wrote a must-read book in the 1920s, when Einstein's theory was first introduced, entitled, "Kings Dethroned." He alone took on the great mathematical mistakes, misrepresentations, and purposely laid out misdirection's from Copernicus to Einstein. Toward the end of his life he ended up having to preach his geocentric truth atop a soap box in his native Carolina hometown when no one in government, academia, or science would listen to him about Flat Earth Theory, or how the Kings of Astronomy had got it all so wrong.

We also must give much credit to the Arab Astronomers of the past who have been completely left out of the Flat Earth topic and conversation. Our basic math, like algebra and calculus was derived from Arab mathematicians who used these equations to better understand the movement of the heavens above. They include the work of Al-Battani (circa 929 AD). His work includes timing of the new moons, calculation of the length of the solar and sidereal year, the prediction of eclipses and the phenomenon of parallax. Al-Sufi (903-986) made several observations on the obliquity of the ecliptic and the motion of the sun (or the length of the solar year). He also made observations and descriptions of the stars, setting out his results constellation by constellation, discussing the star's positions, their magnitudes and their color. Jabir Ibn Aflah (d. 1145) was the first to design a portable celestial sphere to measure and explain the movements of celestial objects. Jabir is specially noted for his work on spherical trigonometry.

Once a generation has been sold such a gradually reinforced lie from country to country, school by school, and across the entire world over decades, the very next generation will believe it as accepted truth without question or investigation. This is how the universal heliocentric lie was sold over the past 6–8 generations (150 + years). One of the biggest questions one asks when debunking heliocentric theory is "why?" Why create such a very long-term, perpetrated hoax about our common cosmology? What does it matter if we go around the Sun or not, or if the shape of the Earth is flat or round?

The simplest answer is that they are afraid of us! Afraid that we will all wake up to our highest of human potentiality. Afraid we will come to know once again, our extreme importance as Earth being the centerpiece of our universe. They are afraid that if we self-realize our highest potential and destiny, we could destroy their thousands of years' empire in a nanosecond and the walls of the Roman Empire would come tumbling down. This is why the Flat Earth discussion is so important to discuss and analyze. Each individual holds power so vast that we all have potential to destroy any power structure in place that does not serve the Cosmic Truths of Nature. This is also why the powers that must cease to be only operate in the dark shadows while using deception, deceit, lies and aliases. They are in constant fear that we as humans come to realize our true power and importance in this Creation.

If we all knew, and deeply understood, that we lived in an enclosed dome, would we continue to abuse Earth and pollute without conscience or consequence?

Would we keep treating Mother Gaia as exclusively as a sewer and source of product to market for sale and profit from if we knew that no one was going to come to save us and what we do to our world has a profound effect on our entire universal being?

We do not need to be lorded over with rules by rulers we never meet or see in our lifetime. They make laws that bind us to their system of governance and resort to using enforcement and coercion to pay tithe on through taxing our labor.

Their 5,000-year reign is now over as the walls and baal Earth come-a-tumbling down. With each awakening to our Flat Earth Theocosmolgy by more and more independent thinkers and researchers we reconnect to our true purpose and meaning as humans on this flat and stationary Earth. We are also reconnecting to our ancient ancestral heritages who possessed such great wisdom and knowledge. We, the Awakening Ones, are just beginning to once again, 'know thyself,' as the inscription above the ancient site in Greece at the Oracle of Delphi directs us. The blinds have been lifted for all to see that Flat, is where it's at!

Onward, Flat Earth Nation, Onward.

– Michael Tellinger, author of "Slave Species of God?" and "Temples of the African Gods"

"After many months of research, I still cannot find any scientifically accepted proof of a physical nature, that the Earth is moving through space – and that it actually is a ball. Everything I held dear and as an unshakable truth has been shaken to its core and stripped of all credibility. It is more clear than ever before that everything we have been told is a lie, and the most important teachings have been strategically omitted by our education system – whose objective is to create a future labour force to keep the money monster alive – and not to teach us to think and give us real skills.

This Flat Earth could be the biggest lie of them all.
Those who think that this will go away soon have a big surprise coming. We have to discard most of NASA's imagery as part of the LIE – which is one of the first disturbing things everyone finds when entering this realm of research.

My most recent research into the relationship between sound, magnetism and electricity clearly shows that all the physical manifestations of shape and matter and the magnetic fields are the same TOROIDAL shapes that are proposed for the sub-atomic model of the electron AND the magnetic fields of the Earth and even the model of the galaxies that seem to spew forth matter at its galactic equator of the gigantic TORUS shaped galaxy.

My breakthrough discovery is that the evidence suggests that the land or Earth itself follows this DOUBLE TORUS model and the land which we have been told is a ball – is actually the FLAT Accretion disk emerging at the centre of the Earth Torus Magnetic field. All North – South Magnetic alignments and everything else that has been attributed to a ball Earth can be explained even better with a FLAT Earth model.
Like a giant RING Magnet with all the expected magnetic fields around it and inside it. This also accounts for the so-called DOME and hollow Earth theory or Agartha, magnetic drift, precessional wobble, Aurora Borealis, and so many more global mysteries. Everything can be explained by simply re-evaluating the magnetic fields around the Earth and the full shape they hold.

For those that thought the Flat Earth Theory is a bunch of ignorant nonsense devised to confuse humanity and not worthy of debate, have a growing scientific mountain of evidence to deal with. I look forward to seeing how this research evolves.

As so many others have done – I have also done my own experiments with the curvature of the Earth and I found to my surprise that THERE IS NO CURVATURE. No matter how hard I looked. And if there is no curvature, there is only one conclusion we can reach. The Earth surface is flat.

If the weather man can show Chicago from Michigan City some 40 odd miles across Lake Michigan, that should be enough proof that there is no curvature. At that distance Chicago should be about 1060 feet below the horizon. And so the new exiting journey begins – sifting through layers upon layers of deception and lies. The only thing I have to hold on to is my sanity, sense of humour and an open mind."

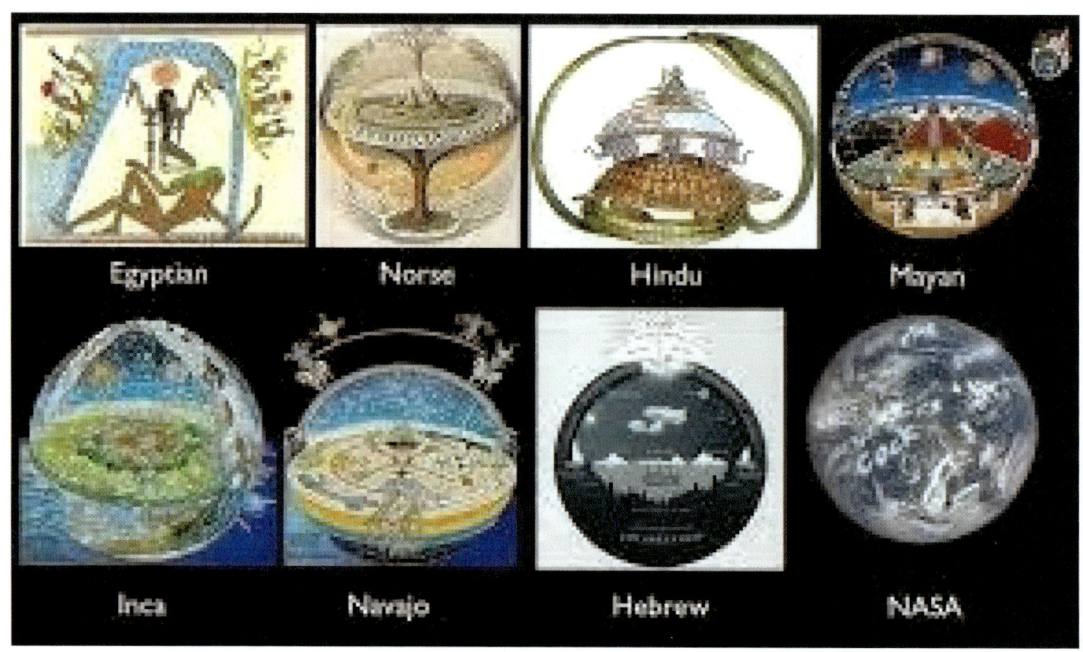

Chapter 1

5,000 Years of Flat Earth TheoCosmology

"Maybe the hero's journey has taken us far enough, and the time has come for a different mythic imagination to rise and offer multiple approaches to the many dilemmas and complicated problems in our World today. Whereas the hero's journey tends to be conceived as a courageous search in distant lands, the genius myth involves a turn within that leads to a sense of self, but also a return to the origins of our lives."
—Michael Meade

Egyptian Cosmology

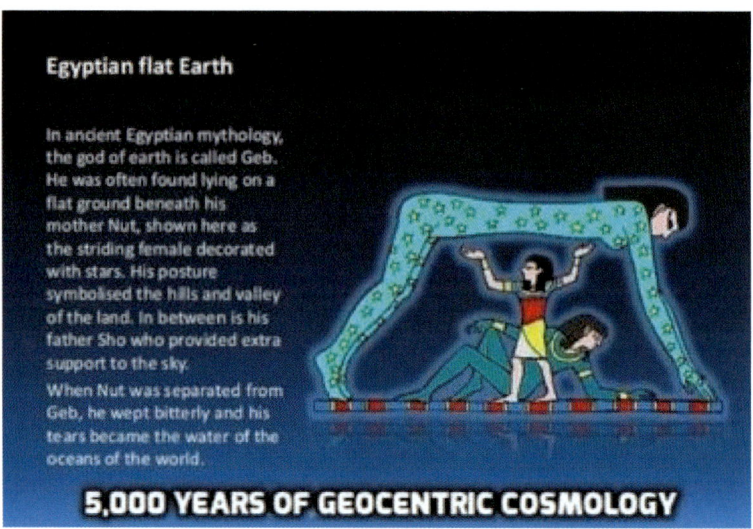

The ancient Egyptians believed the ground was more or less a flat disk that floated upon Nun, the primordial ocean from which Creation had begun in time immemorial and out of which the Nile continued to flow.

The sky was a solid, dome-shaped thing and was often said to be the belly of a goddess—typically *Nut* or *Hathor*, whose body arched over the Earth, with her feet on one horizon and her hands on the other. Her midsection was held aloft by the air god, *Shu*. The stars glimmered on this goddess' belly.

Beneath the surface of the Earth, either within the waters of Nun or just above them, was the underworld, Duat. Duat was where the sun and accordingly the sun god, who was by turns imagined to be Ra, Atum, Khepri, or Amun, went when he sank below the horizon in the evening. During the night, he travelled through the underworld in his boat before emerging with the dawn to continue his journey across the heavens.

Osiris ruled the underworld. When humans went to the underworld after death, part of them would typically make the sun god's circular journey with him and part of them would linger in the underworld with Osiris. During the night, the two parts would be united. The land of Egypt was thought to lie at the center of the cosmos both from a horizontal standpoint and from a vertical one. Each temple occupied the precise center of the cosmos, the axis mundi, for those who worshiped there.

The ancient Egyptian word 'Maat' has been variously translated as 'order,' 'truth,' 'justice,' and 'virtue.' Maat was all of these things and more. It referred to the sacred order and balance of the cosmos. Maat was present in the regular, recurring rhythms of the world—the seasonal rise and fall of the Nile, the daily circle of the sun, etc., as well as in all proper actions. It was thus a concept that described the nature of reality, as well as an ethical concept that guided behavior. Maat and the cosmos were held together by the actions of the gods, and of people, both bound to the timeless standard of rightness.

Egyptian gods and goddesses were each associated with particular forces within what we today would call: 'nature'—the sun, the Nile, the air, the soil, etc.,—which manifested some part of that deity. Not everything in 'nature' was considered divine, only those forces that upheld and sustained the regular cosmic rhythms. Nothing could be assumed to take place mechanically, without the intervention of some active force, because the most basic elements of the world consisted of active forces. Thus, something that modern people would consider as mundane and predictable as a sunrise was cause for ecstatic celebration amongst the ancient Egyptians. If the gods were to leave the world entirely, its underlying structures would be thrown into chaos, and everything that depended on those structures would perish.

Ancient Vedic Cosmology

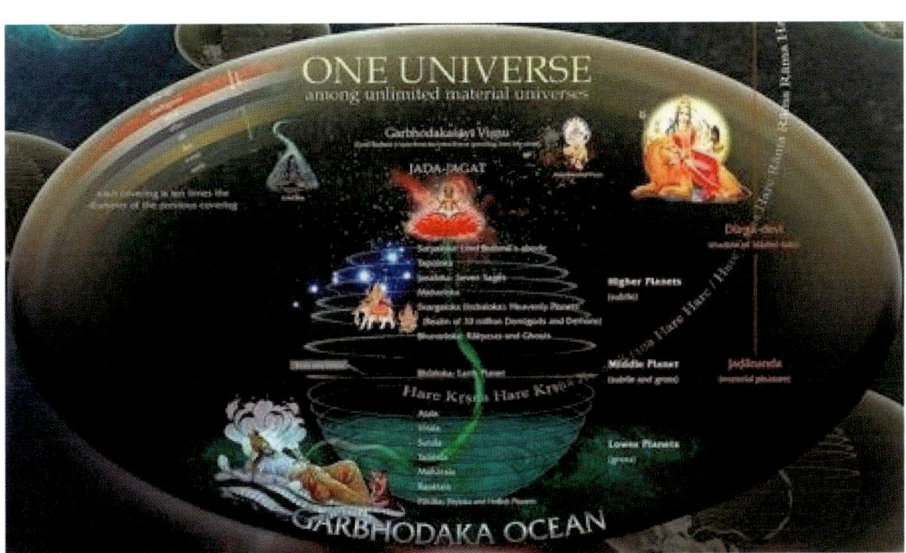

Vedic myth describes the galactic centre. The Vedas also speak of a 'submarine fire' which, …."consumeth the waters of the great Ocean, [and which] became like unto a large horse's head which persons conversant with the Vedas call by the name of Vadavamukha and emitting itself from that mouth it consumeth the waters of the mighty ocean." "Hirto, son of Neph, born of an egg, descended out of the highest heaven. He was a most gracious Lord, and in deference to Om, smote against the rocks of heaven. So, when the egg rocks of heaven. So, when the egg was broken, one-half of the shell ascended, the other half became the foundation of the world."

The Lords Fifth Book

"The primeval God transformed himself into a golden egg which was shining like the sun and in which he himself, Brahman, the father of all worlds, was born. He rested a whole year in this egg and then he parted it into two parts through a mere word. From the two shells, he formed heaven and Earth, in the middle he put the air, and the eight directions of the world, and the eternal dwelling of the water." Creation of the cosmos in Manu. "And out of those two halves he formed heaven and Earth, between them the middle sphere, the eight points of the horizon, and the eternal abode of the waters."

The Laws of Manu

1. Aditya (the sun) is Brahman. In the beginning this was non-existent. It became existent, it grew. It turned into an egg. The egg lay for the time of a year. The egg broke open. The two halves were one of silver, the other of gold.
2. The silver one became this Earth, the golden one the sky, the thick membrane (of the white) the mountains, the thin membrane (of the yoke) the mist with the clouds, the small veins the rivers, the fluid the sea.
3. And what was born from it that was Aditya, the sun. When he was born shouts of hurrah arose, and all beings arose, and all things which they desired. Therefore, whenever the sun rises and sets, shouts of hurrah arise, and all beings arise, and all things which they desire.

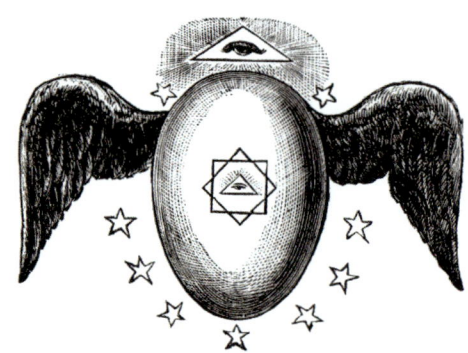

It should be noted here that there is an abundance of cosmic egg theory in the ancient world and even western occultists and alchemists have used it in many symbols—even placing it with a serpent and 'above the stars.' *The Churning of the Milky Ocean* (Milky Way) is a famous and sacred teaching from the Hindu text, *Mahabharata*. To churn the ocean, they used the Serpent King, *Vasuki*, for their churning-string. For a churning-pole they use Mount Mandara, placed on the back of a Great Tortoise—the Kurma Avatar of Vishnu. As the gods and demons churned the sea, a terrible poison issued out of its depths which enveloped the universe.

Nordic Cosmology, the Nine Worlds

The Nine Worlds (old Norse Níu Heimar) are the homelands of the various types of beings found in the pre-Christian world-view of the Norse and other Germanic peoples. They're held in the branches and roots of the world-tree Yggdrasil, the Tree of Knowledge and Wisdom. The Nine Worlds as a group are mentioned in a poem in the Poetic Edda. However, no source gives a list of exactly which worlds comprise the nine. Based on the kinds of beings found in Norse mythology and the reference to their homelands in various literary sources, however, we can compile the following tentative reconstruction:

- **Midgard**–the world of humanity
- **Asgard**–the world of the Aesir tribe of gods and goddesses
- **Vanaheim**–the world of the Vanir tribe of gods and goddesses
- **Jotunheim**–the world of the giants
- **Niflheim**–the primordial world of ice
- **Muspelhiem**–the primordial world of fire
- **Alfheim**–the world of the elves
- **Svartalfheim**–the world of the dwarves
- **Hel**–the world of the eponymous goddess Hel and the dead

With the exception of *Midgard*, they are all primarily invisible worlds, although, in keeping with the animistic and pantheistic character of pre-Christian Germanic religion, they tend to manifest in particular aspects of the visible world. For example, *Jotunheim* overlaps with the physical wilderness, *Hel* with the grave (the literal 'underworld' beneath the ground) and *Asgard* with the sky. While we don't know exactly what the spiritual or magical significance of the number nine was, it's clear that this number had a significance for the pre-Christian Germanic peoples. Philologist Rudolf Simek offers the following summary:

Nine is the mythical number of the Germanic tribes. Documentation for the significance of the number nine is found in both myth and cult. In Odin's self-sacrifice he hung for nine nights on the windy tree (*Hávamál*) there are nine worlds to *Niflhel* (*Vafþrúðnismál 43*), *Heimdallr* was born to nine mothers (*Hyndluljóð 35*) *Freyr* had to wait for nine nights for his marriage to *Gerd* (*Skírnismál 41*) and eight nights (= nine days?) was the time of betrothal given also in the *Þrymskviða*. Literary embellishments in the Eddas similarly use the number nine. *Skaoi* and *Njoror* lived alternately for nine days in *Nóatún* and in *Þrymheimr*; every ninth night, eight equally heavy rings drip from the ring *Draupnir*; *Menglöð* has nine maidens to serve her (*Fjólsvinnsmál 35ff*) and *Aegir* had as many daughters. Thor can take nine steps at the *Ragnarok* after his battle with the Midgard serpent before he falls down dead. Sacrificial feasts lasting nine days are mentioned for both Uppsala and Lejre and at these supposedly nine victims were sacrificed each day.

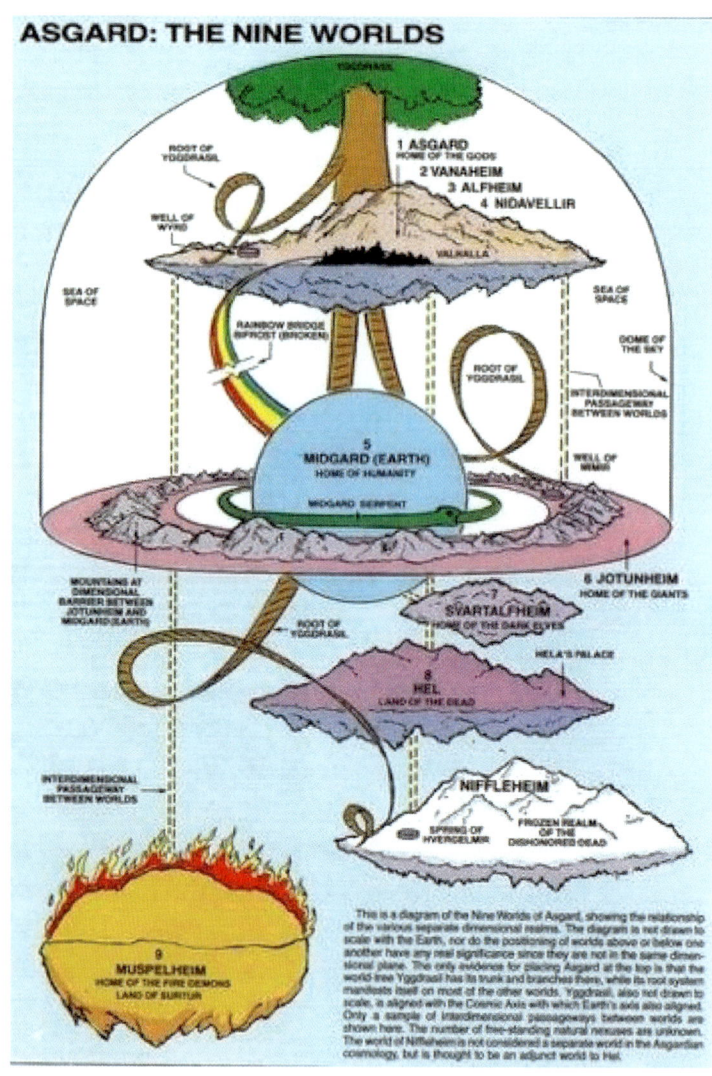

Tibetan Cosmology

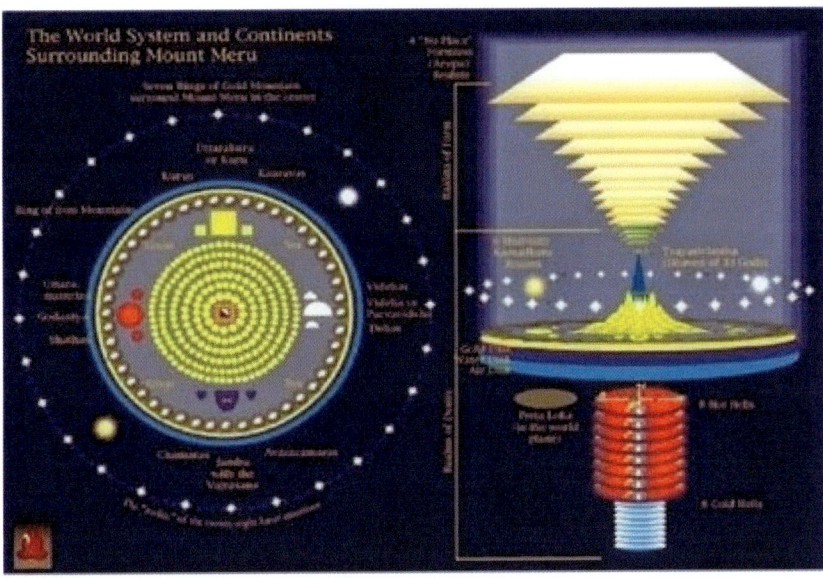

Ancient Buddhists imagined the universe as essentially flat with Mount Meru at the center of all things. Surrounding this universe was a vast expanse of water and surrounding the water was a vast expanse of wind. The universe was made of thirty-one planes of existence, stacked in layers with three realms or dhatus. The three realms were *Ārūpyadhātu*, the formless realm: *Rūpadhātu*, the realm of form and *Kāmadhātu*, the realm of desire. Each were further divided into multiple worlds that were the homes of many sorts of beings. This cosmos was thought to be one of a succession of universes coming into and going out of existence through infinite time. Our world was thought to be a wedge-shaped island continent in a vast sea south of Mount Meru, called *Jambudvipa*, in the realm of *Kāmadhātu*. The Earth, then, was thought to be flat and surrounded by ocean. In Tibetan astro-science, two distinct flat-Earth, stationary, geocentric cosmologies are recognized, both developed in India and later translated into Tibetan. The first is the Abhidharma system, expounded in the 4th or 5th century Indian text Abhidharmakosha (Treasury House of Knowledge) by Vasubandhu, and the Kalachakra system (Wheels of Time) whose root text was translated into Tibetan in 1027 AD. Both systems are mandala-like world systems made of concentric oceans and mountain ranges centered on an axis, Mount Meru. The known world exists on one of the four major continents (with other minor accompanying continents). The southern continent was called Jambudvipa. Mount Meru is lapis-blue on our side, which explains why it cannot be seen but instead blends in with the sky's color.

The heavenly bodies orbit around Mount Meru. When motivated by the accumulated karma of all the sentient beings that existed in the previous world system, the space particles begin the process of forming a new universe from the reformed elements. First air particles coagulate to form wind, which then causes the fire particles to join and create lightning. From this process water particles form rain and the resulting rainbows herald the joining of Earth particles.

The order of this 'creation' is mirrored in the structure of the world system in the Abhidharma and Kalachakra systems depicted above with the air disk being the lowest layer of the base of the world, topped in succession by disks of fire, water, and Earth. Sentient beings begin to populate all possible realms (*e.g.* various hell realms, animal, human, and deities).

During this stage, humans are granted an 'infinite' lifespan (until the end of the age). When this process is complete, the Stage of Formation ends. The Stage of Abiding is subdivided into many eras, in which the lifespan of human's change. We are currently said to be in a degenerate age, when the lifespan is decreasing and is about 80 years. The Stage of Destruction begins when no more sentient beings are reborn in the hell realms, and the hells begin to empty out (as various beings exhaust the karma that brought them there in the first place). Likewise, all higher realms of existence are emptied out in sequence. **Finally, the world system is destroyed by a great fire!**

Far Eastern Cosmology

The ancient Chinese accepted the Earth being a flat geocentric and that an umbrella-like covering surrounded it which was similar to the beliefs of the Greek, East Indian, Egyptians, Norse, and Germanic as well as the aboriginal people of the Americas. The Chinese astronomers believed that heaven takes its body from the yang so it is round and in motion and the Earth takes its body from the Yin, so it is flat and quiescent. The brilliant astronomers of China held this unchanging belief until the Jesuit missionaries came to China in the 17th century.

In Ancient Chinese, the early Jesuit scholars were the first Europeans to gain access to the Chinese 'book of all knowledge' from ancient times. This 4,320-volume collection told of the repercussions of mankind's rebellion against the gods: "The Earth was shaken to its foundations. The sky sank lower towards the North. The sun, moon, and stars changed their motions. The Earth fell to pieces and the waters in its bosom rushed upwards with violence and overflowed the Earth."

The legend of *Nvwa Mending the Sky* is one of the most representative myths about the world creation in ancient China. Legend has it that in remote antiquity, the four pillars supporting the sky suddenly broke and as a result, the sky had chasms and couldn't fully cover the Earth.

The land on Earth also 'spilt' open and couldn't sustain things on top of it. At that time, raging flames ripped through the Earth, where torrential flood water, ferocious beasts and birds also preyed on people's lives. Since human beings were suffering such disasters, *Nvwa* tried everything she could to mend the sky. She selected all types of five-colored stones, melted them into slurry over fire and filled the chasms with the slurry.

Later, she cut off the feet of a big turtle to prop up the sky. Afterwards, she killed the black dragon that caused floods and other disasters and the ferocious beasts and birds harming people. At last, she burnt reeds into ashes and piled them up to block flood water. Thus, disasters were conquered and mankind was saved.

Native American Cosmology

The universe of the Maya was centered on a tiered pyramid, and rest upon a crocodilian cosmic sea. Each quarter of the Earth was associated with color, and the center of the Earth was a 'fifth direction.' Four sacred beings supported the dome of heaven illustrated as a two-headed dragon, whose celestial body was banded with symbols. It was arched over the moon goddess, who was holding the rabbit depicted in the moon's face and a skeletal Venus and the sun god. Pleiades, is a star cluster that has a rattlesnake tail. Creation of both sun and probably the planet Venus was explained with a legend of the Hero twins who vied with the Lords of Death during a series of ball games.

The victorious twins became these celestial bodies. The king of the Inca believed he was the son of the sun. The cosmos was centered on the sun Temple at Cuzco, Peru. In one origin myth the Inca people came from three caves; in another myth they arose from Lake Titicaca. The straight red lines are *ceques*, symbolizing connections to sacred places. The major *ceques* formed borders of the four-quartered Inca world.

The Milky Way blended into the underworld and brought dark, fertile mud to the sky. Upon its return, it formed patches that resembled animals, like the snake (at top) toad, *tinamou* bird, mother and baby llama, fox and a second *tinamou*. The sun is portrayed as a male god and the moon as a female.

To the Cherokees, the Earth was a flat disc of water with a large island floating in the middle. The Earth hung by four cords—one each in the North, East, South, and West—from a sky-arch made of stone. The Middle World was where the plants, animals, and humans lived. Above the sky-arch was the Upper World, which was where the guiding and protective spirits of humans and animals lived.

These spirits could move from the Upper World to the Middle World and back to help the humans keep balance and harmony on the Earth. Below the Earth was the Under World of bad spirits. Bad spirits brought disorder and disaster.

They could rise to the Middle World through deep springs, lakes, and caves. When these spirits caused trouble, Cherokees called on the spirits from the Upper World to help restore the Earth. Everything in the Cherokee environment—from corn, tobacco, and animals to fire, smoke, creeks, and mountains—had an intelligent spirit and played a central role in Cherokee myths and daily practices. Native peoples did not view themselves as separate from their environment—they were a part of it.

In the Aztec creation story, as the gods continued to create, they had a problem—their creations would fall into the water and be eaten by the dreadful *Cipactli*. at the time of war—the four gods attacked the sea monster, pulling her in four directions. She fought back, biting *Tezcatlipoca*, tearing off his foot. At last *Cipactli* was destroyed. From this enormous creature, the universe was created.

All the 13 heavens stretch into her head. The Earth was created in the middle and her tail reaches down to the underworld (*Mictlán*) (nine underworlds, to be exact). You could say that in the Aztec creation story the world is on the back of this sea monster, floating in the water of space (reminiscent of the Iroquois belief that the world rests on the back of a turtle.)

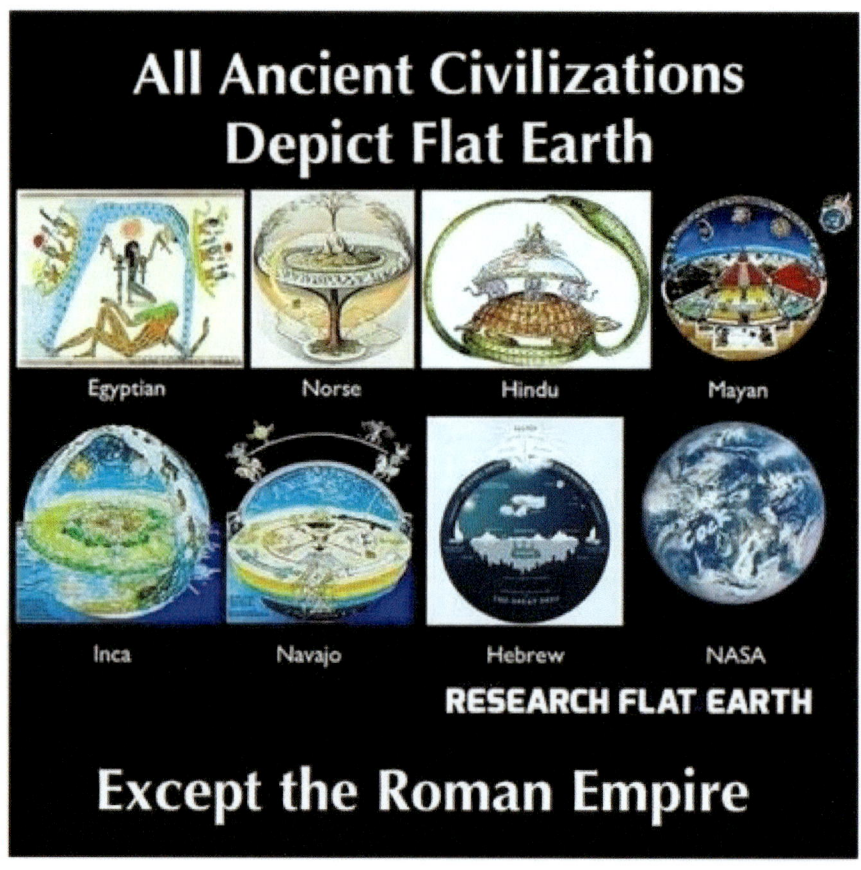

Chapter 2

The Flat Earth Bible and Qur'an

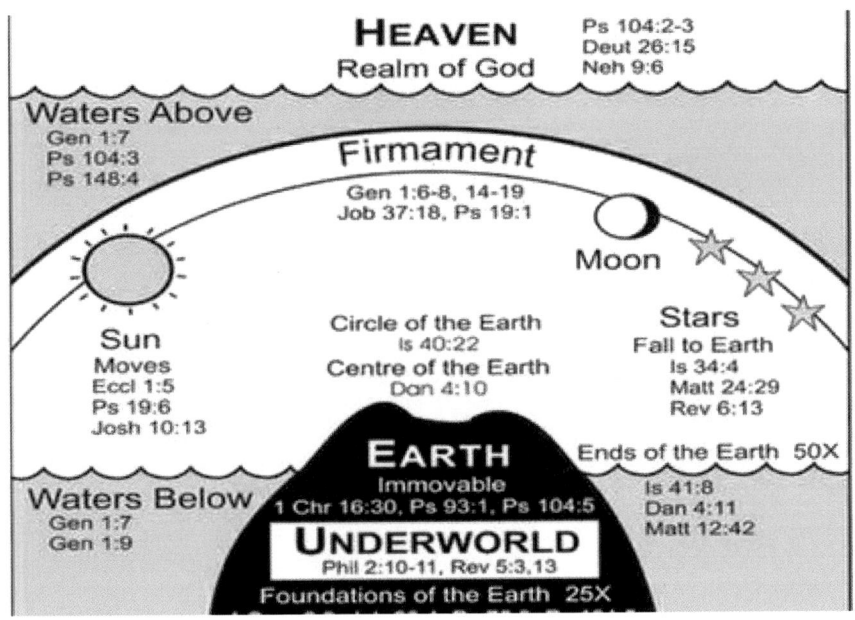

"There is more to human existence and to reality itself than current science can ever give us access to."

—Dalai Lama XIV, The Universe in a Single Atom: The Convergence of Science and Spirituality

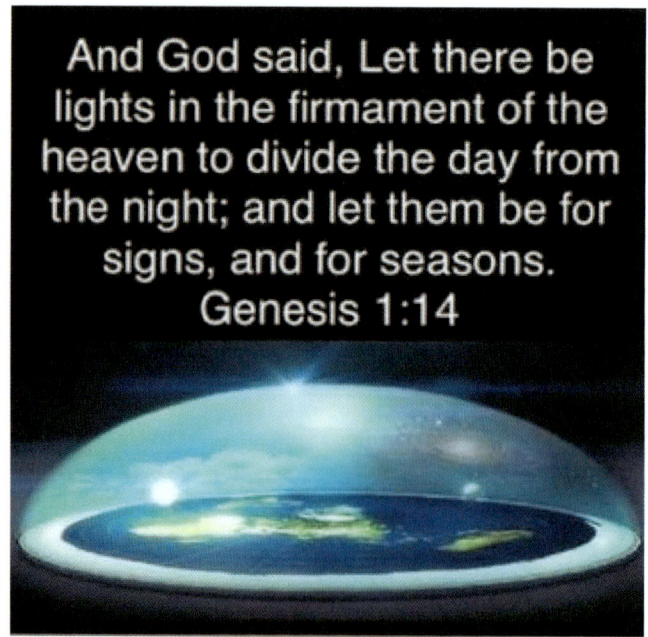

The Firmament as the Dome of Heaven

—Excerpt from God, Reason and the Evangelicals by N.F. Gier

There is more than just poetry in the biblical creation account. In what follows I argue that we should take the Hebrew cosmology as a pre-scientific attempt to understand the universe. Parallel accounts in other ancient mythologies will be the principal evidence I offer. One of the first problems we have is that there is no word in the Hebrew language for the Greek word kosmos. Kosmos was first used by Pythagoras, who is said to be the first Greek to conceive of the universe as a rational, unified whole. Such a notion is crucial to the scientific idea that things operate according to law-like regularity. For the Hebrews, the universe is not a kosmos, but a loose aggregate held together and directed by God's will. If God's will is free—this is an assumption threatened in some evangelical doctrines of God—then the results of such a will are not predictable events, which is why the biblical idea of creation can never be called 'scientific' and why 'scientific creationism' will always be a contradiction in terms.

The most striking feature of the Old Testament world is the 'firmament,' a solid dome which separates "the waters from the waters" (Gen. 1:6). The Hebrew word translated in the Latin Vulgate as *firmamentum* is raqia' whose verb form means 'to spread, stamp, or beat out.' The material beaten out is not directly specified but both biblical and extra-biblical evidence suggests that it is metal. A verb form of raqia' is used in both of these passages: "And gold leaf was hammered out..." (Ex. 39:3); and "beaten silver is brought from Tarshish" (Jer. 10:9). There are indeed figurative uses of this term. A firmament is part of the first vision of Ezekiel (1:22,26) and the editors of the evangelical Theological Wordbook of the Old Testament cite this as evidence that the Hebrews did not believe in a literal sky-dome. It is clear, however, that Ezekiel's throne chariot is the cosmos in miniature and the use of raqia' most likely refers to a solid canopy (it shines 'like crystal') than to a limited space.

The idea of the dome or vault of heaven is found in many Old Testament books, e.g., "God founds his vault upon the Earth…" (Amos 9:6). The Hebrew word translated as 'vault' is 'aguddah whose verb form means to 'bind, fit, or construct.' Commenting on this verse, Richard S. Cripps states that "here it seems that the 'heavens are 'bound' or fitted into a solid vault, the ends of which are upon the Earth." We have seen that raqia' and 'aguddah, whose referent is obviously the same, mean something very different from the empty spatial expanse that some evangelicals suggest.

In the Anchor Bible translation of Psalm 77:18, Mitchell Dahood has found yet another reference to the dome of heaven, which has been obscured by previous translators. The RSV translates galgal as 'whirlwind,' but Dahood argues that galgal is closely related to the Hebrew gullath (bowl) and gulgolet (skull), which definitely gives the idea of 'something domed or vaulted.' In addition, Dahood points out that the parallelism with tebel, 'Earth,' and 'eres, 'netherworld,' suggests that the psalmist is portraying the tripartite division of the universe–heaven, Earth, and underworld.

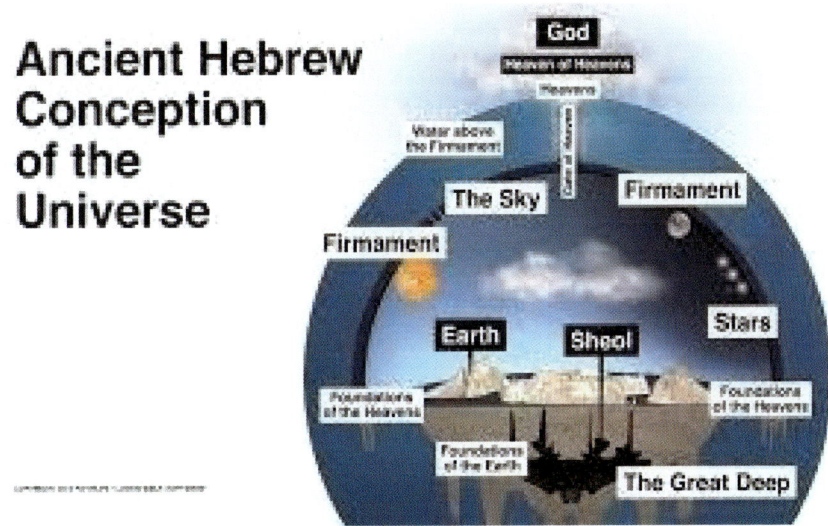

The Pillars of Heaven and Earth

If we disengage ourselves from our own world-view, we can appreciate the internal logic of the Hebrew cosmology. If we are threatened by watery chaos from all sides, then a solid sky would be needed to hold back these ominous seas. If the sky is a solid dome, then it will need pillars to support it. Furthermore, if the Earth is a flat disc floating on "the deep," then it would make sense for it to have some support to hold it in place. One finds the idea of physical supports for heaven in most ancient mythology. One Vedic poet writes of a god "by whom the awesome sky and Earth were made firm, by whom the dome of the sky was propped up";  and Varuna "pillared both the worlds apart as the unborn supported heaven" (Rig-veda 10.121.5; 8.41.10).

The cosmology of the ancient Arabians was a little more advanced. Here we find a solid sky-dome which Allah holds up by an act of will (Surah 2.22). That God "raised up the heavens without pillars" (Surah 13.2) reveals at least two assumptions: (1) that there was something solid to raise up; and (2) earlier views used actual supports and not Allah's direct will.

It is not surprising then that one finds biblical references to the "pillars" or "foundations" of the heaven and Earth. In Job we find that "the pillars of heaven tremble, are astounded at God's rebuke" (26:11). In 2 Samuel we also find that God's anger makes "the foundations of the heavens tremble" (22:8). God's fury also affects the pillars of the Earth: "Who shakes the Earth out of its place, and its pillars tremble?" (Job 9:6); and "the foundations of the world were laid bare at thy rebuke, O Lord, at the blast of the breath of thy nostrils" (Ps. 18:15). There seems to be a little confusion about where the pillars of heaven are located. Generally, in the Bible and other ancient literatures, distant mountains were the most likely candidates. But in one passage at least we find that Yahweh has "laid the beams of his heavenly chambers on the waters" (Ps. 104:3), i.e., the watery chaos surrounding the flat disc of the Earth.

In the Old Testament God is portrayed as a cosmic architect. Isaiah asks: "Who has measured the waters in the hollow of his hand and marked off the heavens with a span?" (40:12). In Proverbs Yahweh "drew a circle on the face of the deep... and marked out the foundations of the Earth..." (8:27-29). God challenges Job with the famous question: "Where were you when I laid the foundations of the Earth? Who determined its measurements... or who stretched the line upon it? On what were its bases sunk, or who laid its cornerstone..." (38:4)? Continuing the same theme, the psalmists ask: "Who placed the Earth upon its foundations lest it should ever quake?" (Ps. 104:5, AB); and observe that "when the Earth totters... it is God who will steady its pillars" (Ps. 75:3, AB). Finally, in 1 Sam. 2:8 we find that "the pillars of the Earth are the Lord's and on them he has set the world." The logic of such a cosmology is expressed well by a Vedic poet: "Water is up there beyond the sky; the sky supports it" (Aitareya Upanishad I.2).

The Waters Above and Below

The Ice Wall surrounds 95% of the Antarctic coast

In her new translation of the *Rig-veda*, Wendy O'Flaherty says that the ancient Hindus believed that "the Earth was spread upon the cosmic waters" and that these primeval oceans "surrounded heaven and Earth, separating the dwelling-place of men and gods…" After the sky fell in on the Celts, the next event they feared was that the seas would come rushing in from all directions. In the Babylonian creation epic *Enuma Elish*, the sky is made from the body of Tiamat, the goddess of watery chaos. The victorious god Marduk splits "her like a shellfish into two parts: half of her he set up and ceiled it as sky, pulled down the bar and posted guards. He bade them to allow not her waters to escape."

In Genesis 1:1 we find the linguistic equivalent of Tiamat in the Hebrew word *tehom* ("the deep"), and the threat of watery chaos is ever present in the Old Testament. Evangelical F. F. Bruce agrees that "*tehom* is probably cognate with Tiamat," and Clark Pinnock admits that Yahweh also "quite plainly…fought with a sea monster" and that the model of the battle is a Babylonian one. The psalmists describe it in graphic terms: "By thy power thou didst cleave the sea-monster in two, and broke the dragon's heads above the waters; thou didst crush the many-headed Leviathan, and threw him to the sharks for food" (Ps. 74:13-14 NEB; cf. Job 3:8; Isa. 27:1).

The firmament separates the waters from the waters, so that there is water above the heavens (Ps. 148:4) and water below the Earth. The Second Commandment makes this clear: "You shall not make for yourself a graven image, or any likeness of anything that is in heaven above, or that is on the Earth beneath, or that is in the water under the Earth…" (Deut. 5:8; cf. Ex. 20:4; Is. 51:6). The lower tier of this three-story universe is identified as water in other passages: "God spread out the Earth upon the waters" (Ps. 136:6); and "he has founded it upon the seas and established it upon the rivers" (Ps. 24:2).

If the waters below the Earth are simply springs, then one would have a hard time making sense of the prohibition of making images of the mostly microscopic creatures found in such waters. The biblical authors are definitely thinking of the great fishes and monsters of "the deep" itself. The fertility goddesses of the land and the seas were Yahweh's principal rivals.

Some evangelicals claim that the author of Job believed that the Earth was suspended in empty space: "The shades below tremble, the waters and their inhabitants. Sheol is naked before God. He stretches out the North over the void, and hangs the Earth upon nothing" (26:5-7). The first thing that can be said here is that the context is not one of God's creation (which comes next at vv. 10-14 following the cosmology above), but one of God's threat of destruction. Second, none of the ancients, except for possibly the Greek atomists, had any notion of empty space. The Hebrew words for "void" and "nothingness" have parallel uses in many Old Testament passages and generally refer to a watery chaos (Gen. 1:1; Jer. 4:23; Is. 40:17, 23). Therefore we must conclude, as does Marvin H. Pope, that Job does not have the Pythagorean notion of the Earth suspended in space. Oceans, not empty space, surround the Hebrew world.

Celestial Chambers and the Heaven of Havens

While it is true that the Hebrews had a rough understanding of the circulation of water vapor and the source of rain in the circulation of water vapor and the source of rain in the clouds (Job 36:27, 28), they also conceived of mechanisms in heaven whereby God could directly induce great atmospheric catastrophes.

Obviously, the clouds themselves could not have held enough water for the Great Flood, so "all the foundations of the great deep burst forth, and the windows of the heavens were opened" (Gen. 7:11; cf. Mal. 3:10). This is also further proof that the Earth was surrounded by watery chaos.

The Old Testament talks about divine 'chambers' (*heder*) in heaven and this notion seems to have been borrowed from Canaanite mythology. Marvin Pope has discovered a direct parallel to the *Ugaritic* god *'El* who 'answers from the seven chambers,' usually through the media of the seven winds. Significantly, we find that Yahweh 'brings forth the wind from his storehouses' (Ps. 135:7); and from the chamber comes the tempest, from the scatter-winds the cold (Job 37:9, AB).

From Amos we learn that God builds his upper chambers in the heavens (9:6), and the psalmists speak of God storing 'his upper chambers' with water so that he can water the mountains (Ps. 104:3, 13; cf. Ps. 33:7). Job gives us the most detailed account of God's chambers: "Have you entered the storehouses of the snow, or have you seen the storehouses of the hail, which I have reserved for the time of trouble, for the day of battle and war?" (38:22).

We must not forget that Yahweh is a warrior (Ex. 15:3) and it is he, for example, who caused the violent storm which destroyed the Canaanite army of Sisera (Jdgs. 5). In the non-canonical Ecclesiasticus we discover that Yahweh has more than storms in his chambers: "In his storehouses, kept for proper time, are fire, famine, disease" (39:29). Dillow argues convincingly that Yahweh's storehouses of rain are not just clouds or ocean basins; rather, they most definitely have a celestial location.

In the diagram above the area above the 'ocean of heaven' is labeled the 'heaven of fire.' I have not been able to verify this and it seems that it must be labeled 'heaven of heavens' instead.

Various levels of heaven are not unique to the Hebrews for we can read that the Vedic seer conceived of at least three superior realms of heaven (Rig-veda 8.41.9). One psalmist clearly distinguishes between the two levels: "You highest heavens, and you waters above the heavens" (Ps. 148:4). This area is exclusively Yahweh's domain: "The heaven of heavens belongs to Yahweh..." (Ps. 115:16, AB); "To the Lord your God belong heaven and the heaven of heavens..." (Deut. 10:14); and "heaven and highest heaven cannot contain thee" (1 Kgs. 8:27).

These passages have led to endless speculation about the various levels of heaven. Creationist Henry Morris claims that there are three heavens: (1) atmospheric heaven (Jer. 4:25); (2) sidereal heaven (Is. 13:10); (3) and the heaven of God's throne (Heb. 9:24). (40) The heaven of heavens mentioned above is probably not Morris' third heaven, because it was created (Ps. 148:4) and it seems that God does not dwell there (1 Kgs. 8:27). Commentators will probably never be able to sort out many of these obscure passages.

Flat Earth Bible Citations

"The whole point of the Copernican theory is to get rid of Jesus by saying there is no up and no down, the spinning ball thing just makes the whole Bible a big joke."
—Charles Johnson

Unless otherwise noted, all the Scriptures are taken from the King James Version. The Scriptures below are arranged by topics. The book of Job is placed before Genesis as many (if not most) scholars believe it is the oldest book of the Bible. This is by no means an exhaustive list of Scripture.

Scriptures concerning the nature of the heavens/sky above and their relationship to the Earth:

Job 9:8
Which alone spreadeth out the heavens, and treadeth upon the waves of the sea. Job 22:14 (HCSB
Clouds veil Him so that He cannot see, as He walks on the circle of the sky.
Job 37:18
Hast thou with him spread out the sky, which is strong, and as a molten looking glass?
Genesis 1:
In the beginning God created the heaven and the Earth.
And the Earth was without form, and void; and darkness was upon the face of the deep. And the Spirit of God moved upon the face of the waters.
And God said, Let there be light: and there was light.
And God saw the light, that it was good: and God divided the light from the darkness.
And God called the light Day, and the darkness he called Night. And the evening and the morning were the first day.
And God said, Let there be a firmament in the midst of the waters, and let it divide the waters from the waters.
And God made the firmament, and divided the waters which were under the firmament from the waters which were above the firmament: and it was so.
And God called the firmament Heaven. And the evening and the morning were the second day.
Psalm 19:1
The heavens declare the glory of God; and the firmament sheweth his handywork.

Psalm 104:
Bless the Lord, O my soul. O Lord my God, thou art very great; thou art clothed with honour and majesty.
Who covereth thyself with light as with a garment: who stretchest out the heavens like a curtain:
Who layeth the beams of his chambers in the waters: who maketh the clouds his chariot: who walketh upon the wings of the wind:

Proverbs 8:27 (ESV)
When He established the heavens, I was there, When He inscribed a circle on the face of the deep.

Isaiah 40:22
It is he that sitteth upon the circle of the Earth, and the inhabitants
thereof are as grasshoppers; that stretcheth out the heavens as a curtain, and spreadeth them out as a tent to dwell in:

Isaiah 44:24
Thus saith the Lord, thy redeemer, and he that formed thee from the womb, I am the Lord that maketh all things; that stretcheth forth the heavens alone; that spreadeth abroad the Earth by myself;

Isaiah 45:12
I have made the Earth, and created man upon it: I, even my hands, have stretched out the heavens, and all their host have I commanded.

Isaiah 48:13
Mine hand also hath laid the foundation of the Earth, and my right hand hath spanned the heavens: when I call unto them, they stand up together.

Isaiah 66:1
Thus saith the Lord, The heaven is my throne, and the Earth is my footstool: where is the house that ye build unto me? And where is the place of my rest?

Ezekiel 1:26
And above the firmament that was over their heads was the likeness of a throne, as the appearance of a sapphire stone: and upon the likeness of the throne was the likeness as the appearance of a man above upon it.

Amos 9:6 (NASB)
The One who builds His upper chambers in the heavens And has founded His vaulted dome over the Earth, He who calls for the waters of the sea And pours them out on the face of the Earth, The Lord is His name. Scriptures concerning the nature of the Earth below the firmament:

Job 26:7
He stretcheth out the north over the empty place, and hangeth the Earth upon nothing.

Job 38:
Then the Lord answered Job out of the whirlwind, and said,
Who is this that darkeneth counsel by words without knowledge?
Gird up now thy loins like a man; for I will demand of thee, and answer thou me.
Where wast thou when I laid the foundations of the Earth? Declare, if thou hast understanding.
Who hath laid the measures thereof, if thou knowest? Or who hath stretched the line upon it?
Whereupon are the foundations thereof fastened? Or who laid the corner stone thereof;
When the morning stars sang together, and all the sons of God shouted for joy?

Job 9:6
Which shaketh the Earth out of her place, and the pillars thereof tremble.

Job 26:10
He hath compassed the waters with bounds, until the day and night come to an end.

Job 28:24
For he looketh to the ends of the Earth, and seeth under the whole heaven;

Job 37:3
He directeth it under the whole heaven, and his lightning unto the ends of the Earth.

Job 38:13
That it might take hold of the ends of the Earth, that the wicked might be shaken out of it?

Genesis 1:
And God said, Let the waters under the heaven be gathered together unto one place, and let the dry land appear: and it was so.
And God called the dry land Earth; and the gathering together of the waters called he Seas: and God saw that it was good.

And God said, Let the Earth bring forth grass, the herb yielding seed, and the fruit tree yielding fruit after his kind, whose seed is in itself, upon the Earth: and it was so.
And the Earth brought forth grass, and herb yielding seed after his kind, and the tree yielding fruit, whose seed was in itself, after his kind: and God saw that it was good.
And the evening and the morning were the third day.
Samuel 2:8
He raiseth up the poor out of the dust, and lifteth up the beggar from the dunghill, to set them among princes, and to make them inherit the throne of glory: for the pillars of the Earth are the Lord's, and he hath set the world upon them.
Samuel 22:16
And the channels of the sea appeared, the foundations of the world were discovered, at the rebuking of the Lord, at the blast of the breath of his nostrils.
1 Chronicles 16:30
Fear before him, all the Earth: the world also shall be stable, that it be not moved.
Psalm 18:15
Then the channels of waters were seen, and the foundations of the world were discovered at thy rebuke, O Lord, at the blast of the breath of thy nostrils.
Psalm 102:25
Of old hast thou laid the foundation of the Earth: and the heavens are the work of thy hands.
Psalm 93:1
The Lord reigneth, he is clothed with majesty; the Lord is clothed with strength, wherewith he hath girded himself: the world also is stablished, that it cannot be moved.
Psalm 96:
O worship the Lord in the beauty of holiness: fear before him, all the Earth.
Say among the heathen that the Lord reigneth: the world also shall be established that it shall not be moved: he shall judge the people righteously.
Let the heavens rejoice, and let the Earth be glad; let the sea roar, and the fullness thereof.
Psalm 104:
Who laid the foundations of the Earth, that it should not be removed forever?
Thou coverest it with the deep as with a garment: the waters stood above the mountains.
At thy rebuke they fled; at the voice of thy thunder they hasted away.
They go up by the mountains; they go down by the valleys unto the place which thou hast founded for them.
Thou hast set a bound that they may not pass over; that they turn not again to cover the Earth.
Psalm 136:
1. To him that stretched out the Earth above the waters: for his mercy endureth for ever.
Proverbs 8:
When he prepared the heavens, I was there: when he set a compass upon the face of the depth:
When he established the clouds above: when he strengthened the fountains of the deep:
When he gave to the sea his decree, that the waters should not pass his commandment: when he appointed the foundations of the Earth:
Isaiah 11:12
And he shall set up an ensign for the nations, and shall assemble the outcasts of Israel, and gather together the dispersed of Judah from the four corners of the Earth.
Isaiah 40:22
It is he that sitteth upon the circle of the Earth, and the inhabitants thereof are as grasshoppers; that stretcheth out the heavens as a curtain, and spreadeth them out as a tent to dwell in:
Isaiah 43:6
I'll say to the north, 'Give them up'! and to the south, 'Don't keep them back!' Bring my sons from far away and my daughters from the ends of the Earth—
Daniel 4:
11. The tree grew, and was strong, and the height thereof reached unto heaven, and the sight thereof to the end of all the Earth:
Matthew 4:
Again, the devil taketh him up into an exceeding high mountain, and sheweth him all the kingdoms of the world, and the glory of them;

9. And saith unto him, All these things will I give thee, if thou wilt fall down and worship me.

Matthew 24:31
And he shall send his angels with a great sound of a trumpet, and they shall gather together his elect from the four winds, from one end of heaven to the other.

John 17:24
Father, I desire that they also, whom thou hast given me, may be with me where I am, to behold my glory which thou hast given me in thy love for me before the foundation of the world.

Revelation 1:7
Behold, he cometh with clouds; and every eye shall see him, and they also which pierced him: and all kindreds of the Earth shall wail because of him. Even so, Amen.

Revelation 7:1
And after these things I saw four angels standing on the four corners of the Earth, holding the four winds of the Earth, that the wind should not blow on the Earth, nor on the sea, nor on any tree.

Revelation 20:8
He will go out to deceive Gog and Magog, the nations at the four corners of the Earth, and gather them for war. They are as numerous as the sands of the seashore. Scriptures concerning the nature of the sun, moon, and stars.

Genesis 1:
And God said, Let there be lights in the firmament of the heaven to divide the day from the night; and let them be for signs, and for seasons, and for days, and years:

And let them be for lights in the firmament of the heaven to give light upon the Earth: and it was so.

And God made two great lights; the greater light to rule the day, and the lesser light to rule the night: he made the stars also.

And God set them in the firmament of the heaven to give light upon the Earth,

And to rule over the day and over the night, and to divide the light from the darkness: and God saw that it was good.

Psalms 19:
The heavens declare the glory of God; and the firmament sheweth his handywork.

Day unto day uttereth speech, and night unto night sheweth knowledge.

There is no speech nor language, where their voice is not heard.

Their line is gone out through all the Earth, and their words to the end of the world. In them hath he set a tabernacle for the sun,

Which is as a bridegroom coming out of his chamber, and rejoiceth as a strong man to run a race.

His going forth is from the end of the heaven, and his circuit unto the ends of it: and there is nothing hid from the heat thereof.

Psalm 136:
8. To him that made great lights: for his mercy endureth for ever:
9. The sun to rule by day: for his mercy endureth for ever:
10. The moon and stars to rule by night: for his mercy endureth for ever.

Ecclesiastes 1:5
The sun also ariseth, and the sun goeth down, and hasteth to his place where he arose.

Joshua 10:13 And the sun stood still, and the moon stayed, until the people had avenged themselves upon their enemies. Is not this written in the book of Jasher? So the sun stood still in the midst of heaven, and hasted not to go down about a whole day.

Isaiah 38:8
Behold, I will bring again the shadow of the degrees, which is gone down in the sun dial of Ahaz, ten degrees backward. So the sun returned ten degrees, by which degrees it was gone down.

Habakkuk 3:11
The sun and moon stood still in their habitation: at the light of thine arrows they went, and at the shining of thy glittering spear.

*****

Whereas very much of what goes by the name of 'science' to-day is not science at all. It is only hypothesis! Read Man's books on this so-called science, and you will get tired of the never-ending repetition of such words as 'hypothesis,' 'conjecture,' 'supposition,' etc., etc.

This is the reason that such theories, which are falsely dignified by the name of science, are constantly changing. We talk of the 'Science of Geology,' or of 'Medical Science'; but read books on geology or medicine, for example, written fifty years ago, and you will find that they are now quite 'out of date.' But truth cannot change.

Truth will never be 'out of date.' What we know can never alter! This of itself proves that the word science is wrongly used when it is applied only to hypotheses, which are merely invented to explain certain phenomena. It is not for such theories that we are going to give up facts. It is not for conjectures that we are going to abandon truth.". W. Bullinger (Number in Scripture, p. 103)

Flat Earth Qur'an

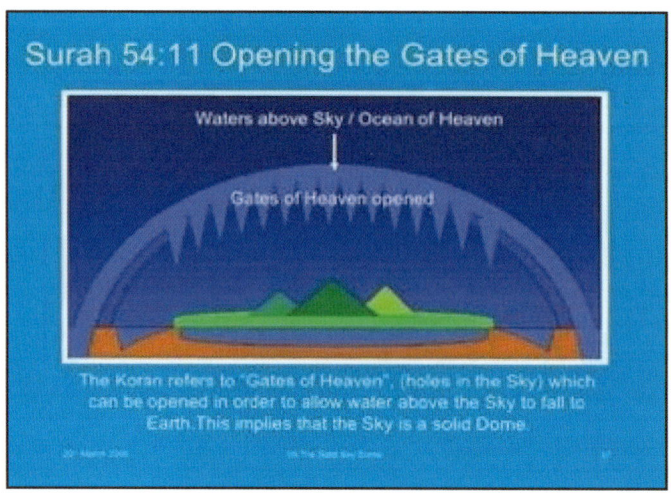

If you accept the literal truth of every word of the Bible, then the Earth must be flat. The same is true for the Qur'an. Pronouncing the Earth round then means you're an atheist. In 1993, the supreme religious authority of Saudi Arabia, Sheik Abdel-Aziz Ibn Baaz, issued an edict, or fatwa, declaring that the world is flat. Anyone of the round persuasion does not believe in God and should be punished.—Carl Sagan

The Qur'an teaches that Allah made the Earth a bed or a carpet which spread out and placed on it mountains as pegs to keep it stable. Allah also made the heavens as a roof, a dome to cover the Earth. The Qur'an further presumes that the Earth is stationary, remaining in a fixed position, not moving.

Allah mentions His perfect ability and infinite authority, since it is He Who has raised the heavens without pillars by His permission and order. He, by His leave, order and power, has elevated the heavens high above the Earth, distant and far away from reach. The heaven nearest to the present world encompasses the Earth from all directions, and is also high above it from every direction. The distance between the first heaven and the Earth is five hundred years from every direction, and its thickness is also five hundred years. The second heaven surrounds the first heaven from every direction, encompassing everything that the latter carries, with a thickness also of five hundred years and a distance between them of five hundred years. The same is also true about the third, the fourth, the fifth, the sixth and the seventh heavens... Then Allah mentions the Earth and how He placed in it mountains standing firm, which make it stable and keep it from shaking in such a manner that the creatures dwelling on it would not be able to live. Hence Allah says ... And the mountains. He has fixed firmly. (79:32)

Whenever `Umar bin Al-Khattab, may Allah be pleased with him, swore an emphatic oath, he would say, "No, by the One by Whose command the heaven and the Earth stand," i.e., THEY STAND FIRM by His command to them and His subjugation of them.

Then, when the Day of Resurrection comes, the Day when the Earth will be exchanged with another Earth and the dead will come forth from their graves, brought back to life by His command and His call to them. The first (thing) created by God is the Pen. It proceeded to (write) whatever is going to be. (God) then lifted up the water vapor, and the heavens were created from it. Then He created the fish, and the Earth was spread out on its back.

The fish moved, with the result that the Earth was shaken up. It was steadied by means of the mountains, for the mountains indeed proudly (tower) over the Earth. So he said, and he recited: "*Nun*. By the Pen and what they write." It was said that '*Nun*' refers to a great whale that rides on the currents of the waters of the great ocean and on its back it carries the seven Earths, as was stated by Imam Abu Jafar Ibn Jarir.

Flat Earth Qur'an Quotes

Qur'an 15:19
And t We have spread out (like a carpet); set thereon mountains firm and immovable; and produced therein all kinds of things in due balance.

Qur'an 20:53
"He Who has, made for you the earth like a carpet spread out; has enabled you to go about therein by roads (and channels); and has sent down water from the sky." With it have We produced diverse pairs of plants each separate from the others.

Qur'an 43:10
(Yea, the same that) has made for you the earth (like a carpet) spread out, and has made for you roads (and channels) therein, in order that ye may find guidance (on the way);

Qur'an 50:7
And the earth- We have spread it out, and set thereon mountains standing firm, and produced therein every kind of beautiful growth (in pairs)-

Qur'an 51:48
And We have spread out the (spacious) earth: How excellently We do spread out!

Qur'an 71:19
And Allah has made the earth for you as a carpet (spread out),
Have We not made the earth as a wide expanse.

"So this modern atheistic, Big Bang, heliocentric globe earth, chance evolution, paradigm spiritually controls humanity by removing God or any sort of intelligent design from the mind and replaces purposeful divine creation with haphazard random cosmic coincidence. And so by removing Earth from the motionless center of the universe these Masons have moved us physically and metaphysically from a place of supreme importance to one of complete nihilistic indifference.

If the Earth is the center of the Universe then the ideas of God creation and a purpose for human existence are resplendent, but if the Earth is just one of billions of planets revolving around billions of stars and billions of galaxies then the ideas of God creations and a specific purpose of for Earth and human existence become highly implausible." — Eric Dubay, The Flat Earth Conspiracy

"The heliocentric theory, by putting the sun at the center of the universe…made man appear to be just one of a possible host of wanderers drifting through a cold sky. It seemed less likely that he was born to live gloriously and to attain paradise upon his death. Less likely, too, was it that he was the object of God's ministrations."—Morris Kline

Chapter 3
Flat Earth Is

The facts are simple. The Earth is Flat.
—Charles K. Johnson,
President of the International Flat Earth Research Society.

The Story of Us
Heliocentric vs. Geocentric Theories

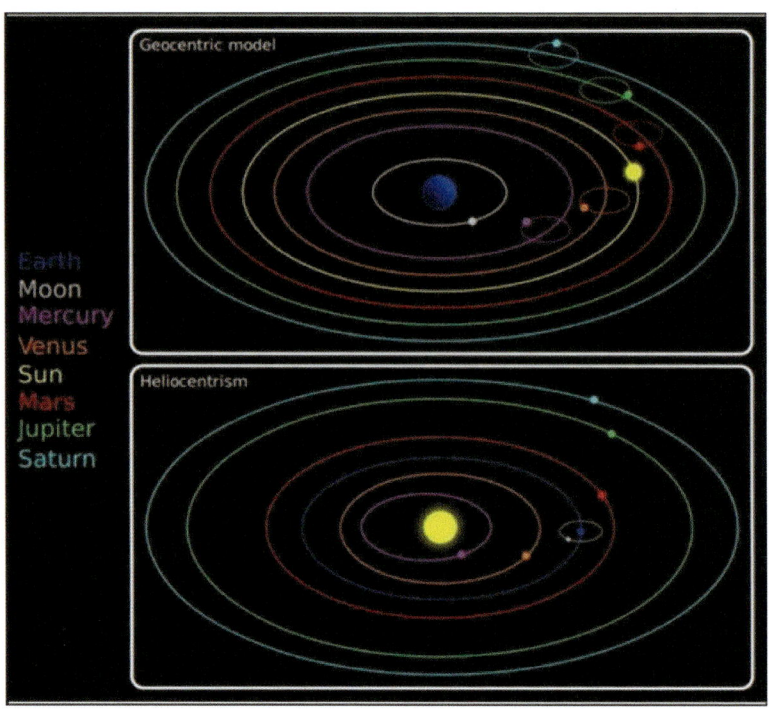

The etymology of the word 'planet' actually comes from late Old English *planete*, from Old French *planete* (modern French planète), from Latin *planeta*, from Greek *planetes*, from (asteres) *planetai* wandering (stars), from *planasthai* 'to wander,' of unknown origin, possibly from PIE *pele 'flat, to spread' or notion of 'spread out.' And Plane (n) 'flat surface,' circa 1600, from Latin *planum* 'flat surface, plane, level, plain,' *planus* 'flat, level, even, plain, clear.' They just added a 't' to our Earth plane and everyone bought it.

In ancient Latin language the letter 'T' represented 'Terra Firma' or Earth. So adding a 'T' to plane, the derived the world 'plane-t"

Figure of the heavenly bodies — An illustration of the Ptolemaic geocentric system by Portuguese cosmographer and cartographer Bartolomeu Velho, 1568 (Bibliothèque Nationale, Paris)

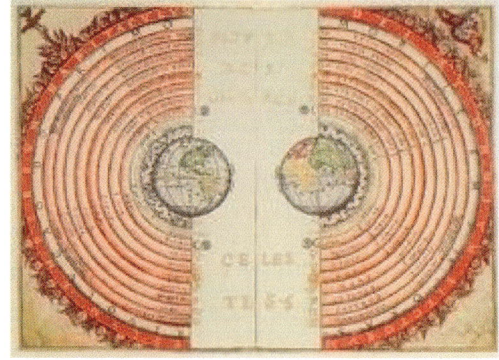

When Flat Earth TheoCosmology Went "Away"
Riccioli's "Almagestrum Nova"

In 1543, Copernicus suggested the sun was at the center of the cosmos. When Jesuit astronomer, Giovanni Battista Riccioli published his Almagestrum Novum or "New Almagest," the title alone suggested the boldness of the project. This was to be a new and updated take on Ptolemy's Almagest. The book offered new insight into the state of thought about the cosmos in 17th century Europe. The frontispiece to Riccioli's Almagestrum Novum tells his perspective on the state of astronomy in 1651. Urania, the winged muse of astronomy, holds up a scale with two competing models, a sun centered Copernican model, and the Tychonic geocentric model.

What this famous depiction about whether the Earth rotated around the Sun or vice-versa was any and all debate about heliocentric round ball Earth vs. geocentric Flat Earth! The Ptolemaic model sits discarded in the bottom right corner of the scene. On the right, 100-eyed Argus points at cherubs in the upper right corner of the illustration. The Cherubs hold recent observational discoveries; the moons of Jupiter, a detailed mountainous moon and the rings of Saturn. Under God's hand from the top of the image, the scale reports the Tychonic model to be heavier and thus the winner.

Some of the most interesting details in this illustration are tucked away in the corners. In the upper right corner, among the clouds, are small representations of additional solar systems, further selling the heliocentric model. Beyond the central diagram, the mapmaker shows the concept of the plurality of worlds. Each of these little sets of circles represents more solar system with a star and planets. This image directly draws on the literary author, de Fontenelle, who building on the ideas of Newton and Descartes', explored the significance of living in a universe with a plurality of worlds each orbiting their own stars.

Riccioli leans on the authority of a number of contemporary and historical thinkers. He lists 38 different astronomers and thinkers, such as Aristotle, Ptolemy and others who believe the Earth to be the center of the universe. He compares them to the 16 astronomers, including Copernicus, Kepler, and Descartes, who favor a sun centered model. With this painting as reference, the Vatican effectively eliminated 5,000 years of geocentric astronomy.

Are We the Center of This Creation?

The earliest recorded example of a geocentric universe comes from around the 6th century BCE. It was during this time that Pre-Socratic philosopher Anaximander proposed a cosmological system where a cylindrical Earth was held aloft at the center of everything. Meanwhile, the Sun, Moon, and planets were holes in invisible wheels surrounding the Earth, through which humans could see concealed fire.

During this same century, the Pythagoreans began to propose that the Earth was circular, based on observation of eclipses (and in all likelihood, observations of the zodiac from different latitudes). By the 4th century BCE, this idea combined with the concept of a geocentric universe to create the cosmological system that most Greeks subscribed to. It was also during the 4th century BCE that Plato and Aristotle would create works on the geocentric universe that would secure its place as the predominant cosmological theory. According to Plato, the Earth was a sphere and the stationary center of the universe. The stars and planets were carried around the Earth on spheres or circles, arranged in the order of distance from the center. These were the Moon, the Sun, Venus, Mercury, Mars, Jupiter, Saturn, fixed stars, and the fixed stars. The ancient called them the "Luminaries."

His system was expanded by Eudoxus of Cnidus, a contemporary of Plato's who developed a less mythical, more mathematical explanation of the luminarie's motion based on the Platonic idea of uniform circular motion. Aristotle elaborated on Eudoxus' system, placing a spherical Earth at the center and all other heavenly bodies arranged in concentric crystalline (i.e. transparent) spheres around it.

Just One of Billions and Billions and Billions or Numero Uno?

There are in fact 100 billion galaxies, each of which contain something like 100 billion stars. Think of how many stars, and planets, and kinds of life there must be in this vast and awesome universe.
~ Carl Sagan, Astronomer and NASA progagandist (1994)

A light-year is how astronomers measure distance in space. It's defined by how far a beam of light travels in one year—a distance of six trillion miles. Throughout the universe, we are told that all light travels at exactly the same speed: about 670 million miles per hour or 186,00 miles per second.

The Milky Way galaxy, in which we're told our Sun and all the stars we see at night reside within, is said to span 100,000 light-years from one end to the other. Astronomers tell us that they can send out a tiny little radio beam and have it bounce off an object up to 587, 862,554,124,841,000 miles or 587 quadrillion miles away. They then can determine its composition, make-up, orbital path, etc.

Putting that into perspective, the duration of recorded modern civilization is about 6,000 years. Again, we have no frame of reference to comprehend such distances. We must take these incredible numbers on faith of scientific say so alone.

According to NASA, our universe spans a distance of approximately 276,000,000,000,000,000,000,000 miles or 276 quintillion miles across and it is said to be expanding further each and every day at the speed of light. This, we are told, has been going on since the Big Bang, some 14.5 billion years ago.

The Big Bang is said to have occurred from nothing into everything our universe is today, and still expanding. Something from nothing came everything. Magicians revel in the grand illusion of it all.

It needs to be clearly understood that a Jesuit priest, Georges Lemaitre, first noted in 1927 that an expanding universe could be traced back in time to an originating single point, scientists have built on his idea of cosmic expansion.

Again, why is Vatican authoring the creation and promotion of the Big Bang Theory if we were created by God?

Commercial Air Routes Match Flat Earth
Not A Round Globe Spinning Ball

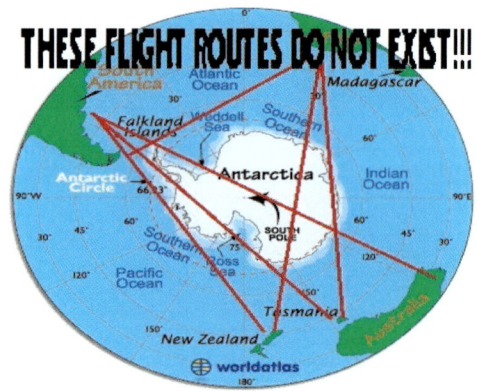

According to the Federal Aviation Association, no air flights have gone the much shorter routes over the Arctic and Antarctic due to cold weather effecting flights as well as having to effect rescue over such inhabited areas (what about the oceans?!). Another reason that is given is that airlines would be required to carry special survival equipment (jackets, boots?), even though thousands of hours of flight time and fuel consumption would be saved by going direct instead of Westward and Eastward directions.

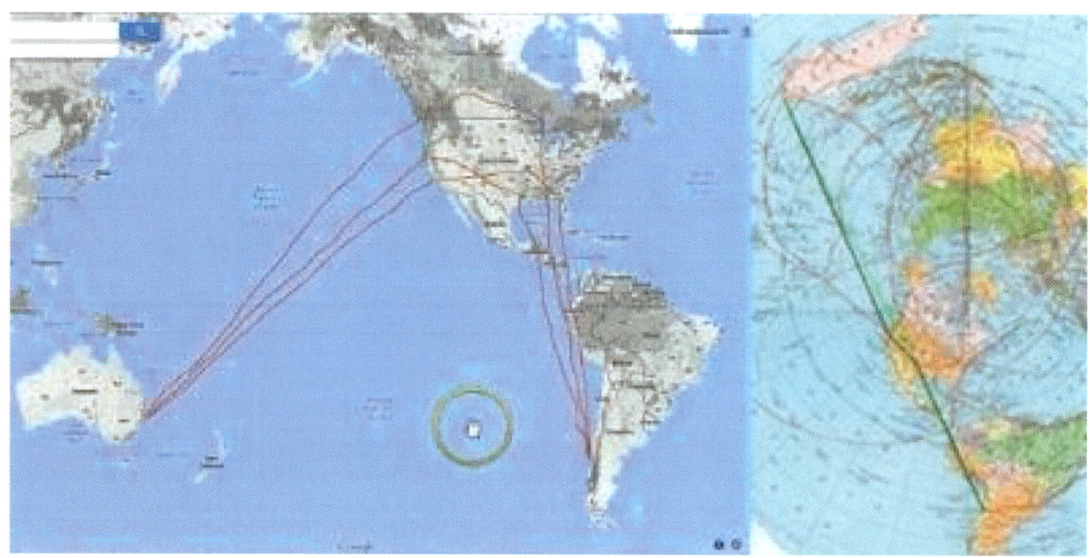

"If Earth was a ball, and Antarctica was too cold to fly over, the only logical way to fly from Sydney to Santiago would be a straight shot over the Pacific staying in the Southern hemisphere the entire way. Refueling could be done in New Zealand or other Southern hemisphere destinations along the way if necessary. In fact, however, Santiago-Sydney flights go into the Northern hemisphere making stop-overs at LAX and other North American airports before continuing back down to the Southern hemisphere. Such ridiculously wayward detours make no sense on the globe but make perfect sense and form nearly straight lines when shown on a flat Earth map." ~ Eric Dubay 200 proofs Earth not a Spinning Ball.

"On a ball-Earth, Johannesburg, South Africa to Perth, Australia should be a straight shot over the Indian Ocean with convenient re-fueling possibilities on Mauritus or Madagascar. In actual practice, however, most Johannesburg to Perth flights curiously stop over either in Dubai, Hong Kong or Malaysia all of which make no sense on the ball, but are completely understandable when mapped on a flat Earth." ~ Ibid

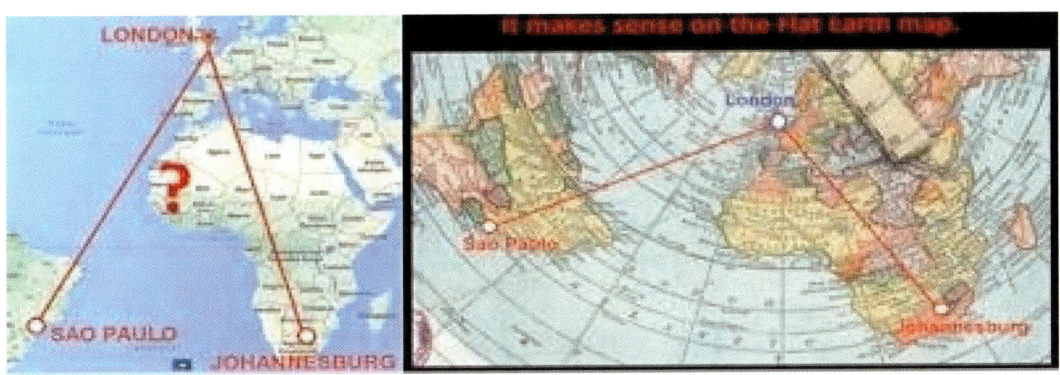

"On a ball-Earth Johannesburg, South Africa to Sao Paolo, Brazil should be a quick straight shot along the 25th Southern latitude, but instead nearly every flight makes a re-fueling stop at the 50th degree North latitude in London first! The only reason such a ridiculous stop-over works in reality is because the Earth is flat." ~ Ibid

They Just Make This Stuff Up
The Awakening Is Now

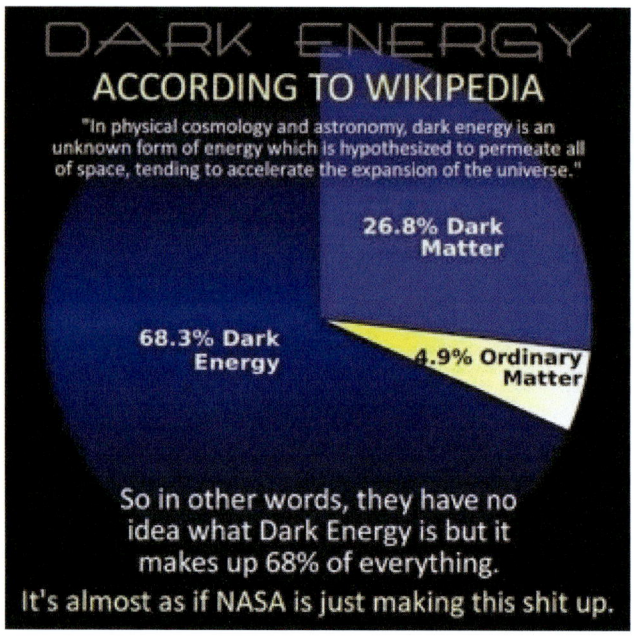

In conventional science, all the "stuff" of the universe fills a regime of cosmic nothingness, with quantum mechanical electromagnetic fluctuations at an extremely small sub-atomic level filling up this "nothingness"—the so-called zero-point energy. Virtual particles supposedly pop in and pop out of existence—unpredictably, chaotically, randomly—to satisfy or not satisfy mass-energy conservation. Recently more baggage has been added to this cosmic picture by conventional science: It feels a need to augment the universe with so far unidentified "dark matter," "dark energy," "quintessence," and a seemingly interminable epicyclic bestiary of imagined creatures to help patch up the Big Bang with its primary structural feature, curved space-time, as dictated by General Relativity. This is Einstein's theory that supposedly "explains" gravity, but which does no such thing.

The powers that control media and all stories from space have had to perpetuate the heliocentric theory with ever more outrageous findings over and over again. They report distances that are beyond human experience to comprehend much less understand the science behind it all. Science without Spiritual connection is amoral and very dangerous. The message they wish to tattoo into all is that we are small and insignificant and live in a small and insignificant not so special corner of a massive galaxy. Our entire faith of who we are is a blind faith in the academic and scientific process of decades of research and advancements in technology. This coupled with an absolute certainty that such a grand hoax could even take place or they could get away with it.

The Big Bang scientific findings are that **"from nothing we became everything."** This is rationally accepted as scientific theory and proof. In addition, the movies and social medias reinforce that we are nothing from nothing. So it must be true!

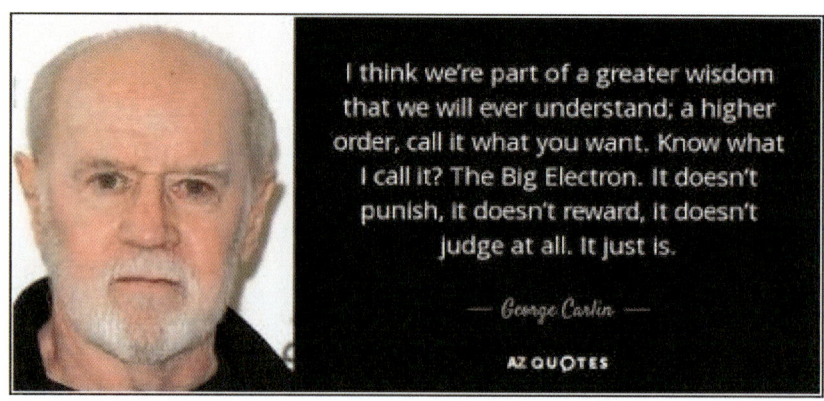

The Sun is Proven NOT to be 93 Million Miles Away

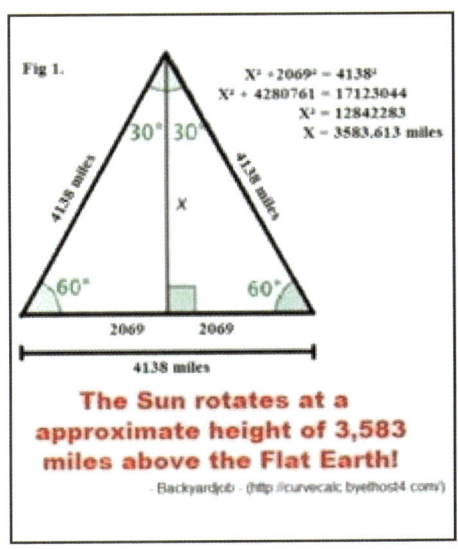
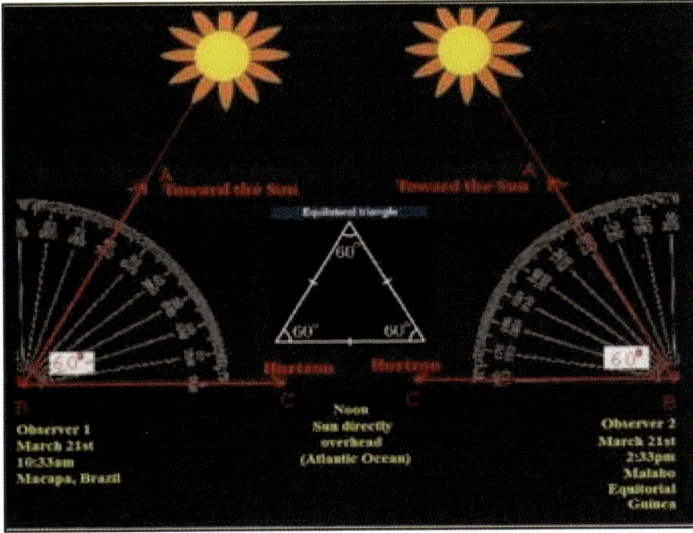

Nicolas Copernicus calculated the sun's distance from Earth to be 3,391,200 miles. The next century Johannes Kepler decided it was actually 12,376,800 miles away. Issac Newton once said, "It matters not whether we reckon it 28 or 54 million miles distant for either would do just as well." This is called modern day science? Accredited scientists like Benjamin Martin calculated between 81 and 82 million miles, Thomas Dilworth claimed 93,726,900 miles, John Hind stated positively 95,298,260 miles, Benjamin Gould said more than 96 million miles and Christian Mayer thought it was more than 104 million.

~ Eric Dubay

Throughout the ages, ancient cosmologists like the Egyptians, Babylonians, and Chaldeans used sextants and plane trigonometry to calculate that the Sun and Moon were both only about 32 miles in diameter and approximately 3,600 miles from Earth. An equilateral triangle has three equal length sides and three 60° angles. If a person is standing in equatorial Macapá, Brazil on the equinox (March 21st at 10:33am) looking east and observes the sun 60° above the horizon and at that exact same moment of time (four time zones away–2:33pm) a person in Malabo.

Equatorial Guinea is looking west and observes the Sun 60° above the horizon then the distance between these two observers is exactly the same as the distance between either observer and the Sun. The distance between the two locations is 4,138 miles or 6,654 kilometers measured by Google Earth. Now using a sextant and recording the angles of the sun from both locations, the distance to the Sun is then known to be precisely 3,583.613 miles above our heads, NOT 93 million miles from Earth.

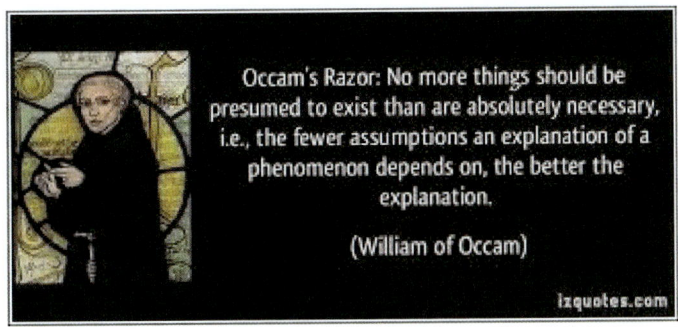

Ancient Sundials Tracked Only on a Flat Earth Model

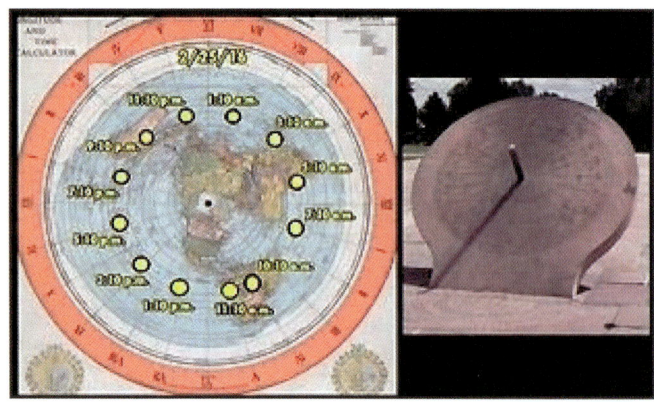 

"Space is a partial vacuum: its different regions are defined by the various atmospheres and 'winds' that dominate within them, and extend to the point at which those winds give way to those beyond. Geospace extends from Earth's atmosphere to the outer reaches of Earth's magnetic field, whereupon it gives way to the solar wind of interplanetary space. Interplanetary space extends to the heliopause, whereupon the solar wind gives way to the winds of the interstellar medium.

Interstellar space then continues to the edges of the galaxy, where it fades into the intergalactic void. Outer space, or simply just space, is the void that exists between celestial bodies, including the Earth. It is not completely empty, but consists of a hard vacuum containing a low density of particles, predominantly a plasma of hydrogen and helium as well as electromagnetic radiation, magnetic fields, neutrinos, dust and cosmic rays." —Wiki, Outer Space

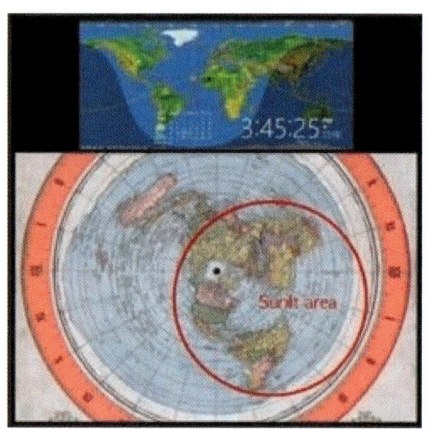

There are huge centuries-old stone sundials and moon dials all over the world which still tell the time now down to the minute as perfectly as the day they were made. If the Earth, Sun and moon were truly subject to the number of contradictory revolving, rotating, wobbling and spiralling motions claimed by modern astronomy, it would be impossible for these monuments to so accurately tell time without constant adjustment.

Sun and Moon Circling
in Concert Overhead in Elegant Simplicity

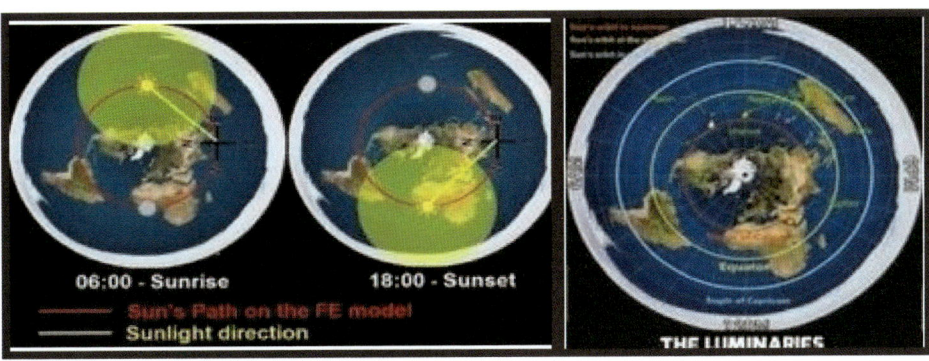

"The earth is a stretched-out structure, which diverges from the central north in all directions towards the south. The equator, being midway between the north center and the southern circumference, divides the course of the sun into north and south declination. The longest circle around the world which the sun makes, is when it has reached its greatest southern declination. Gradually going northwards the circle is contracted. In about three months after the southern extremity of its path has been reached, the sun makes a circle around the equator. Still pursuing a northerly course as it goes around and above the world, in another three months the greatest northern declination is reached, when the sun again begins to go towards the south. In north latitudes, when the sun is going north, it rises earlier each day, is higher at noon and sets later; while in southern latitudes at the same time, the sun as a matter of course rises later, reaches a lesser altitude at noon and sets earlier.

In northern latitudes during the southern summer, say from September to December, the sun rises later each day, is lower at noon and sets earlier; while in the south he rises earlier, reaches a higher altitude at noon, and sets later each day. **This movement around the earth daily is the cause of the alternations of day and night; while his northerly and southerly courses produce the seasons.** When the sun is south of the equator it is summer in the south and winter in the north; and vice versa. The fact of the alternation of the seasons flatly contradicts the Newtonian delusion that the earth revolves in an orbit around the sun. It is said that summer is caused by the earth being nearest the sun, and winter by its being farthest from the sun. But if the reader will follow the argument in any text book he will see that according to the theory, when the earth is nearest the sun there must be summer in both northern and southern latitudes; and in like manner when it is farthest from the sun, it must be winter all over the earth at the same time, because the whole of the globe-earth would then be farthest from the sun!!! In short, it is impossible to account for the recurrence of the seasons on the assumption that the earth is globular and that it revolves in an orbit around the sun."
-Thomas Winship, "Zetetic Cosmogeny" (124-125)

"The seasons are caused by the Sun's circuit around the Earth in a spiral ecliptic. In the Winter Solstice (December 21st), the Sun is vertical over the Tropic of Capricorn. Looking South from London, he appears to make a small circuit in the Southern sky, during the same period he is seen to cross the sky at almost overhead in Cape Town, thus causing Summer in the Southern Hemisphere. In the Summer Solstice (June 21st), the Sun is vertical over the Tropic of Cancer, (nearly overhead in London), while looking North from Cape Town, he appears to make a small circuit in the Northern sky, causing Winter in the Southern and Summer in the Northern Hemisphere." -E. Eschini, "Foundations of Many Generations"

The Law of Perspective

"The theory which affirms that all parallel lines converge to one and the same point on the eye-line, is an error. It is true only of lines equi-distant from the eye-line; lines more or less apart meet the eye-line at different distances, and the point at which they meet is that only where each forms the angle of one minute of a degree, or such other angular measure as may be decided upon as the vanishing point. This is the true law of perspective as shown by nature herself; any idea to the contrary is fallacious, and will deceive whoever may hold and apply it to practice."

"What can be more common than the observation that, standing at one end of a long row of lamp-posts, those nearest to us seem to be the highest; and those farthest away the lowest; whilst, as we move along towards the opposite end of the series, those which we approach seem to get higher, and those we are leaving behind appear to gradually become lower… It is an ordinary effect of perspective for an object to appear lower and lower as the observer goes farther and farther away from it. Let any one try the experiment of looking at a light-house, church spire, monument, gas lamp, or other elevated object, from a distance of only a few yards, and notice the angle at which it is observed. On going farther away, the angle under which it is seen will diminish, and the object will appear lower and lower as the distance of the observer increases, until, at a certain point, the line of sight to the object, and the apparently uprising surface of the earth upon or over which it stands, will converge to the angle which constitutes the 'vanishing point' or the horizon; beyond which it will be invisible." -Dr. Samuel Rowbotham, "Zetetic Astronomy, Earth Not a Globe!" (230-1)

The Sun is Really Moving Across the Daytime Sky
The Sun and Moon are the Same Size

When one observes the sun and moon, we can easily observe for ourselves two equally-sized equidistant circles tracing similar paths at similar speeds above us. The sun is really moving across the sky along with the moon just as anyone can plainly see.

When the Royal Institute of Astronomy and their colleagues, The Society of Jesuits, began to tell/sell the heliocentric model to the masses back in the mid-1800s, they could not convince the peasants, laymen and farmers, who mostly lived and slept outside, that the Sun was not moving. This could be a reason why still to this day we call the Sun's movement a **'sunrise' and 'sunset'**.

One would have thought that with all the astrophysicists, astronomers, academicians, NASA scientists and English majors, someone would have challenged such a gross error in the description of such basic heliocentric physics after all this time, yet they have never done so. Why?

Sidebar: A famous Egyptian mythological narrative tells the story of a Flat earth sunrise and sunset. Osiris, the Sun God ruled ancient Egypt, along with his Goddess wife, *Isis*. They had two sons, Horus and Set. Horus was the good son, who obeyed his parents and became a role model his parents were proud of. The second son, Set, was the proverbial bad boy who dreamed of killing his father and taking over his power one day so he could rule over all. One day, Set abducted his father, Horus, and cut him up into 14 pieces. He then threw all of the cut up pieces of Horus into the Nile river. Isis, being a Goddess, was able to recover all of Osiris's body parts, except for his penis. With the help of the good son, Osiris, they were able to put Horus back together again, except for his missing regenerating organ.

Now fully back in power, Osiris decided to honor his favorite Son (Sun) that had brought light to his day. So he named the Eastern light that brings new life to each of our days, the horizon or the Horus-Son. Set, who had brought darkness and death we now call a Sun-set. The uncapped pyramid with the one-eye on the back of a US dollar bill is also known as the 'eye of Horus' The one that always has an eye on you, as in big brother is always watching you.

Flat Earth, It's About Time Time Zones
Only Work on a Flat Map

Today's scientists have substituted mathematics for experiments, and they wander off through equation after equation, and eventually build a structure which has no relation to reality." —Nikola Tesla

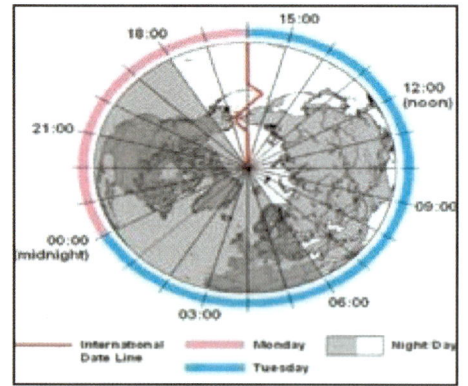
What time is it at the North Pole?

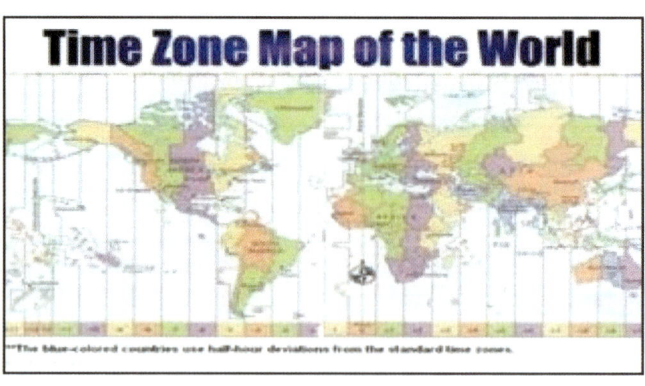
Time zones only work on a Flat map!

According to heliocentric theory, the Earth spins on an imaginary pole called its axis. Every 24 hours, the Earth is said to make one complete rotation we call a day. The days are divided equally in one hour segments, as seen in the flat map to the right.

However, if the earth is a round ball sphere, then necessarily the time zone lines would all converge at the North and South poles and time would shrink as we reached the top or bottom of the poles. Now add that we are told by science that the Earth is spinning at a constant 1,000 mile-per-hour every day and tell me what time is it at the poles on a round ball Earth?

Additionally, you will only see flat maps used for ship navigation, never do they use round ball projection map with adjustments for curvature of the Earth in their calculations. Why?

Crepuscular Rays Prove a Flat Earth
Do you believe them or your own lyin' eyes of observation?

According to modern physicists and astronomers, crepuscular rays, also known as sunbeams, sun rays, and God rays, are rays of sunlight that "*appear* to radiate from the point in the sky where the Sun is located, despite *seeming* to converge at a point." Crepuscular comes from the Latin word "crepusculum," meaning twilight. The words '**apparently**' and '**seemingly**' are used throughout heliocentric theory to explain phenomenon we can clearly see is not true for ourselves.

The picture below shows how sunlight would hit Earth if it were coming from 93 million miles away, just like a flashlight spans our wide over greater distance and focuses like a single beam when up close, just like the photographs you see above.

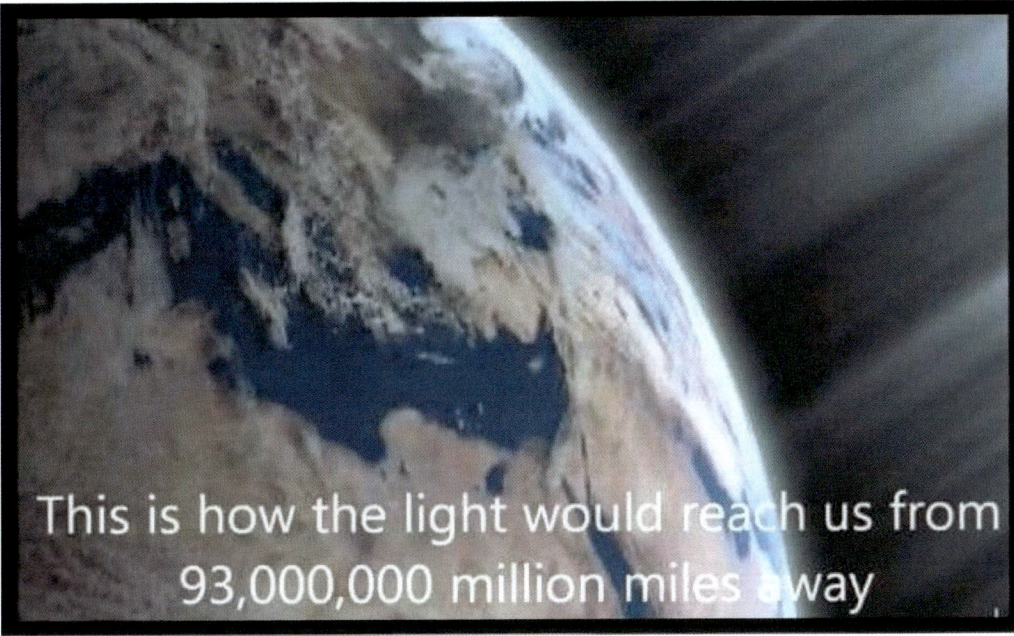

Basic Spherical Geometry
Miles x Miles x 8 inches
Measuring the Curve

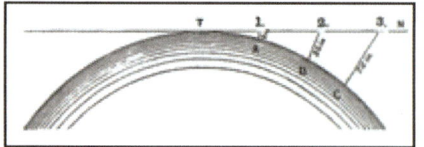

According to basic spherical geometrical mathematics on a curved round ball, the fall off the curve the farther one travels is measured miles x miles x 8 inches.

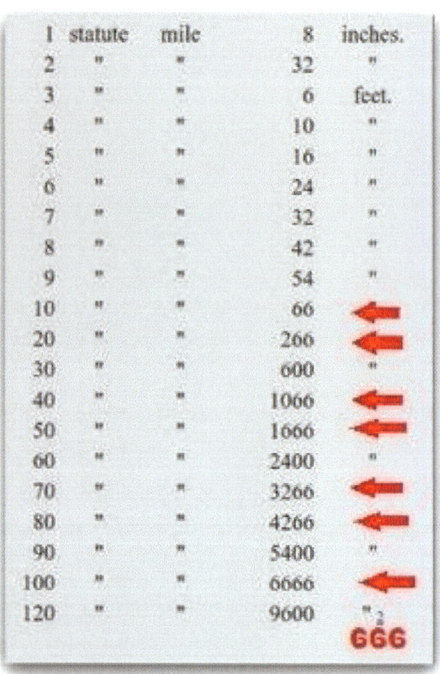

Above is the table that tells how inches and feet one would have to adjust for the curvature on a round ball. Salar de Uyuni is the world's largest salt flat at 4,086 square miles (10,852 square kilometers). The salt flats are located in southwest Bolivia, near the crest of the Andes and is nearly 12,000 ft. above sea level. The Salar was formed as a result of transformations between several prehistoric lakes. It is covered by a few meters of salt crust, which has an extraordinary flatness. The average altitude **varies only within one meter (3 1/4 ft.) over the entire land mass of the Salar.**

The salt flats are 100 miles long by 84 miles wide and **should have a drop of some 1.626 miles in length** if the Earth was a sphere. This proves that we are not living on a round ball globe.

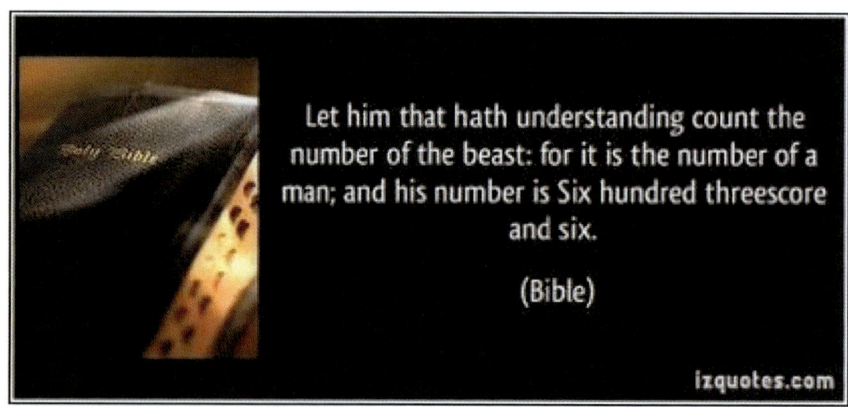

The Ocean Doesn't Curve
The Horizon Always Rises to Meet the Eye, Never Falls Away

"Astronomers are in the habit of considering two points on the Earth's surface, without, it seems, any limit as to the distance that lies between them, as being on a level, and the intervening section, even though it be an ocean, as a vast 'hill'—of water! The Atlantic Ocean, in taking this view of the matter, would form a 'hill of water' more than a hundred miles high! The idea is simply monstrous, and could only be entertained by scientists whose whole business is made up of materials of the same description: and it certainly requires no argument to deduce, from such 'science' as this, a satisfactory proof that the Earth is not a globe.

Every man in full command of his senses knows that a level surface is a flat or horizontal one; but astronomers tell us that the true level is the curved surface of a globe! They know that man requires a level surface on which to live, so they give him one in name which is not one in fact! This is the best that astronomers, with their theoretical science, can do for their fellow creatures–deceive them."

—William Carpenter, "100 Proofs the Earth is Not a Globe"

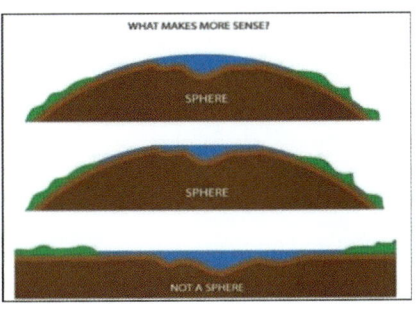

A simple observation can be done the next time you have an unobstructed view, such as on an ocean beach. The average human's ability to see distance looking forward on a clear day with and unobstructed view to the horizon is between 2 ½ to 3 miles. Additionally, looking left-to-right following the width of the or view a distance of 10 miles can be seen. We should see the Earth curve on our perimeters some **33** feet on both sides, according to spherical geometry, and a rounded crest in front of us. The total drop off for ten miles is **66** feet. The water is perfectly flat and does not rise nor fall away from our viewing perspective. Science will tell us that Earth is too big to see the curve. Basic knowledge of spherical geometry and simple self-observation prove this not to be true.

"We'll Be 'Cruising Level' at 30,000 ft....for the Next 5 hours"

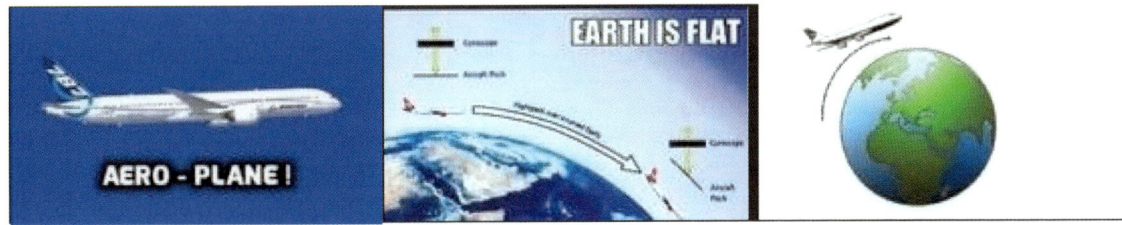

If the Earth were truly a sphere, airplane pilots would have to constantly correct their altitudes nose down as to not fly straight off into outer space! If the Earth were truly a sphere, curving 8 inches per mile squared, a pilot wishing to simply maintain their altitude at a typical cruising speed of 500 mph would have to constantly dip their nose downwards and descend 2,777 feet every minute of flight time yet the curve is not adjusted for and the nose of the plane does not point down to go around the curve.

A plane flying at a typical 35,000 feet wishing to maintain that altitude at the upper-rim of the so-called 'Troposphere' in one hour would find themselves over 200,000 feet high into the 'Mesosphere' without correction for Earth curvature. Without compensation for the curve, in one hour's time, the pilot would find themselves 31.5 miles higher than expected.

Some heliocentrists will argue that still unproven gravity magically keeps aero-*planes* at the perfect altitude around the curve without any need for compensation of travel around a round ball. Earth.

The round ball Earth is said to be spinning from West to East to greet the Sun *rising* each day. If the Earth is spinning at 1,000 mph, West to East and a plane were to take off from California to New York cruising in the same direction as the spinning below, how can an aero-*plane* traveling at only 500 mph, ever reach New York? Conversely, a 5-hour flight traveling from California to Hawaii, against the 1,000 mph spin of the Earth below, should only take less than 1 ½ hours of flight time.

Flights in the Southern Hemisphere to the Northern Hemisphere would also have to compensate for a rapidly spinning ball. If leaving Brazil, flying to New York, the pilot would have to direct the plane towards Near Asia to arrive in New York some 10 hours later. Flight times are never, ever calculated, or adjusted for on flight times.

First Ever Photograph of Earth from Space in 1946   Flaaaaaaaaaat!

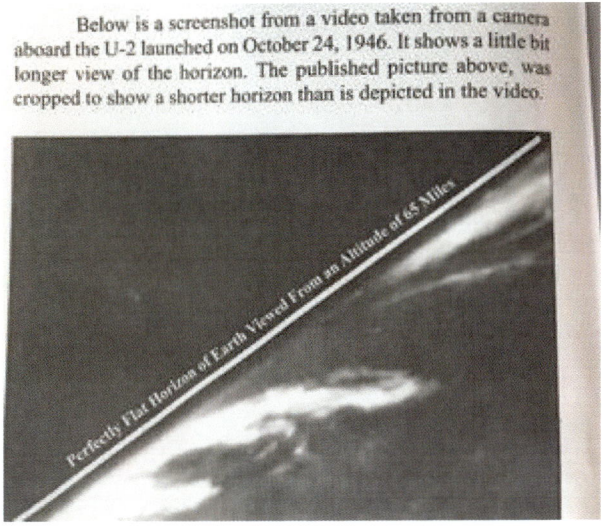

Attitude Horizon Indicator
Gyroscopes Never Adjust for the Curve

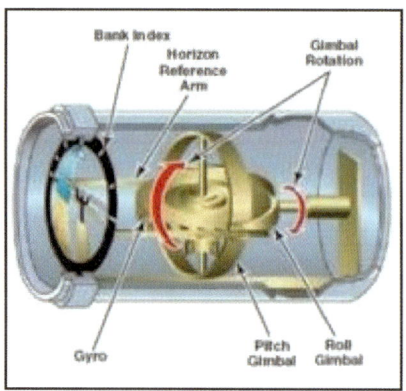 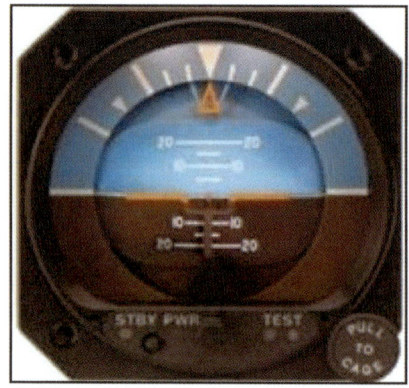

"I hold that space cannot be curved, for the simple reason that it can have no properties. It might as well be said that God has properties. He has not, but only attributes and these are of our own making. Of properties, we can only speak when dealing with matter filling the space. To say that in the presence of large bodies space becomes curved is equivalent to stating that something can act upon nothing. I, for one, refuse to subscribe to such a view."
—Nikola Tesla

An altitude indicator (AI), is also known as gyro horizon or artificial horizon or attitude director indicator (ADI). It is an instrument used in an aircraft to inform the pilot of the orientation of the aircraft relative to Earth's horizon. It indicates pitch (fore and aft tilt) and bank (side to side tilt) and is a primary instrument for flying in weather conditions. The mechanics of the device itself is a mechanical gimbal gyroscope. If you ask an aeronautical pilot, he/she will tell you that the ADI is what they rely on to keep level altitude with Earth while in flight. Most just assume that this instrument calculates and adjusts for the Earth's curvature. When you ask any pilot how a simple Gimbal gyroscope can account for the curve of Earth when flying, they will tell you that it is done electronically, somehow. When informed that the ADI is solely mechanical and does not adjust for curvature, pilots will often then respond, 'well it must be gravity that keeps us locked to Earth's curvature.'

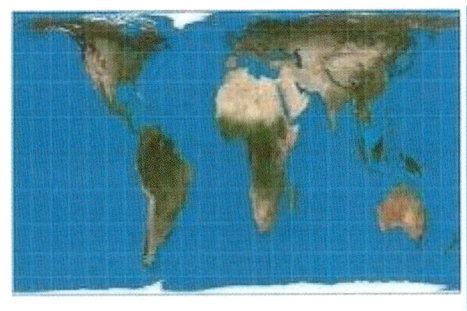 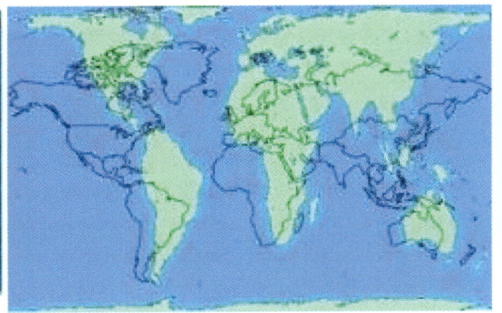 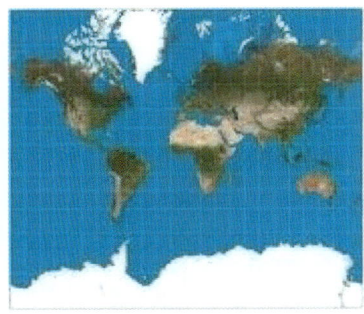

Gall/Peters Projection   *Overlay of both maps*   *Mercator Projection*

World Maps Are Wrong

The Mercator Projection Map is the most common map used in schools, textbooks and relied upon by millions of people every single day. The true scale of land masses relative in size and shape to one another are all wrong and gives us a very distorted representation of the world. The traditional "Mercator Projection Map" has a cultural and ethnic bias because of the relative and distorted sizes of the land masses creating a Eurocentric bias favoring Western civilization and greatly minimizing the size of South America and Africa. The Mercator Map represents Greenland and Africa as being roughly the same size however in reality, Africa is 4 times as large as Greenland. Africa is in fact 30.221 x 1000 sq. Km.

The American Cartographer Association (ACA) calls for an accurate flat, round Earth Map. The ACA, now Cartography and Geographic Information Society, has produced a series of booklets (including *Which Map Is Best*) designed to educate the public about map projections and distortion in maps. In 1989 and 1990, after some internal debate, seven North American geographic organizations adopted the following resolution, **which rejected all rectangular world maps, a category that includes both the Mercator and the Gall–Peters projections**:

WHEREAS, *the earth is round* with a coordinate system composed entirely of circles, and WHEREAS, *flat world maps are more useful than globe maps*, but flattening the globe surface necessarily greatly changes the appearance of Earth's features and coordinate systems, WHEREAS, world maps have a powerful and lasting effect on people's impressions of the shapes and sizes of lands and seas, their arrangement, and the nature of the coordinate system, WHEREAS, frequently seeing a greatly distorted map tends to make it "look right."

THEREFORE, we strongly urge book and map publishers, the media and government agencies **to cease using rectangular world maps** for general purposes or artistic displays. Such maps promote serious, erroneous conceptions by severely distorting large sections of the world, by showing the round Earth as having straight edges and sharp corners, by representing most distances and direct routes incorrectly, and by portraying the circular coordinate system as a squared grid. The most widely displayed rectangular world map is the Mercator (in fact a navigational diagram devised for nautical charts), but other rectangular world maps proposed as replacements for the Mercator also display a greatly distorted image of the spherical Earth.

All Is Frequency, Vibration, and Energy

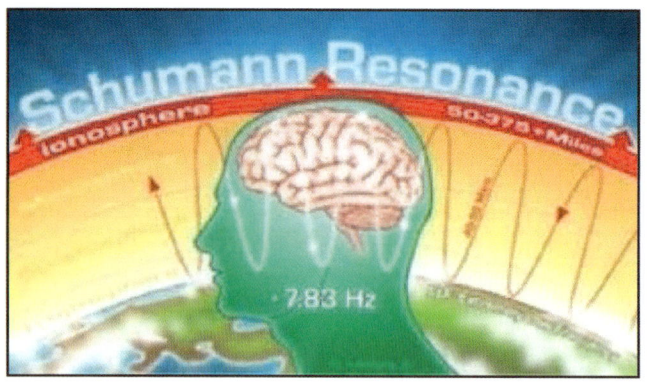

Alpha waves in the human brain are between 6 and 8 hertz. The wave frequency of the human cavity resonates between 6 and 8 hertz. All biological systems operate in the same frequency range. The human brain's alpha waves function in this range and the electrical resonance of the Earth is between 6 and 8 hertz. Thus, our entire biological system—the brain and the Earth itself—work on the same frequencies. If we can control that resonate system electronically, we can directly control the entire mental system of humankind."
—Nikola Tesla

Mr. Tesla noted that 'ether' was everywhere, moving and dynamic, and the salvation of humankind. He also predicted that "with the power derived from it, with every form of energy obtained without effort, from stores forever inexhaustible, humanity will advance with giant strides."

Mr. Tesla was unrivaled for his innate understanding of the Cosmic Laws of Nature that govern all beings and Earth properties. The Aether, or air, as we call it is the existence of electromagnetic fields that interconnect all species of life to the Earth's & Sky's harmonic resonance or frequency also known as the Schumann Frequency.

The ionosphere below the dome surrounding our Earth is electrically and positively (cathode) charged. At the far North pole, a negative charge (anode) occurs. As Earth is a self-contained battery, so are we also electromagnetically charged at nearly the same exact resonate frequencies as the Aether above and all around, and Earth below our feet.

"The day Science begins to study nonphysical phenomena, it will make more progress in one decade than in all previous centuries of its existence."

Mr. Tesla considered that any infringement of cosmic space, or the magnetic field of the Earth, was an infringement of harmony inherent to natural laws. This made him stand up against using the use of atomic energy. In addition, he believed, the absence of ethical components in science has had the consequence of negative influences on people's free will, which makes it destructive. According to Tesla, free will can be creative *only with kindness*, which accompanies the higher understanding and conscious selection of positive intentions. He considered that humankind living on the Earth should understand all kinds of natural alliances with Earth - otherwise people would lose it.

The inside of the ionosphere layer is used in wireless communication to transfer information by bouncing off waves emitted by transmitters on the Earth's surface. In this way, the information can be transferred over large distances. Tesla was the first to carry out wireless energy experiments at Colorado Springs, USA, which produced such powerful electrical tensions that they resulted in the creation of artificial lightning. These lightning flashes also produced radio waves. Due to their extremely low frequency these waves could penetrate the Earth without resistance and thereby Tesla discovered the resonance frequency of the Earth. Unfortunately, Tesla was before his time and his discoveries were not taken seriously.

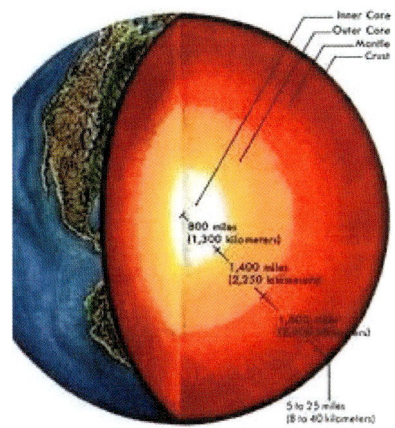

According to heliocentric theory, there are two poles that are activated by a deep iron molted core at Earth's center. This core is said to be as powerful as our Sun and allows Earth's magnetic shield to function and protect us from deadly Sun rays. How can scientists possibly know that the Earth core is made up of Iron when the deepest drilling operation in history, the Russian Kola Ultradeep, managed to get only 8 miles down into the Earth's core? This makes the entire ball- Earth model taught in schools showing a crust, outer-mantle, inner-mantle, outer-core and inner-core layers are all purely speculation as we have never penetrated through beyond the crust.

"Of all the forms of nature's immeasurable, all-pervading energy, which ever and ever change and move, like a soul animates an innate universe, electricity and magnetism are perhaps the most fascinating. .....We know that electricity acts like an in-compressible fluid; that there must be a constant quantity of it in nature; that it can neither be produced or destroyed... and that electricity and ether phenomena are identical."
—Nikola Tesla

The Dome is Our Home
The Tesla Shield

We live under an electromagnetic dome as proven by the Mr. Tesla when he was able to combine Earth with Sky energy to create artificial lightning storms causing damage to land and homes nearby at his laboratory in Colorado Springs, Colorado in 1900.

Tesla himself felt that his work in Colorado Springs would profoundly change our understanding of Earth astrophysics. "It was on the 3rd of July -- the date I shall never forget -- when I obtained the first decisive experimental evidence of a truth of overwhelming importance for the advancement of humanity," Tesla wrote in his journal. In short, as artificial lightning he had created with the help of his laboratory, got farther away, the pulses being picked up by Tesla's equipment didn't fade. Tesla felt he had discovered evidence that the Earth itself contained "stationary waves" that could serve as a good conduit for electromagnetic energy, opening the possibility of worldwide, instantaneous communication and world-wide transmission of power through the Earth's crust.

Tesla and his assistant, Kolman Czito, sent 12 million volts flying through their 80-foot mast, shooting bolts of lightning 145 feet in all directions. The dazzling light display that night ultimately sent Colorado Springs into darkness after a generator operated by the city power company melted down. It took weeks of pleading from Tesla, along with the promise to fix the generator free of charge, before the power company reconnected the juice to Tesla's lab. In one three-hour period, Tesla says he counted upwards of 12,000 lightning strikes in the area detectable by his equipment.

Now, the only public remembrance of Tesla's legacy in Colorado Springs is a small, laminated plaque adorning a historic marker that's tucked into a clump of trees on Memorial Park's northern border along Pikes Peak Avenue. "It was at this facility on North Foote Avenue that Tesla felt he made his most important discoveries," states the plaque, which age has made hard to read through cracks, stains and decay.

Electromagnetism, or magnetic levitation (maglev) is what modern-day trains run on. They can achieve much greater speeds of quieter travel due to Tesla's discovery of magnetic alternating current that produces much less drag and at higher speeds of travel than conventional railroad combustion engines. This is also how the Sun and Moon track in quantum lock step above us all each day. It is the magnetic power of the Moon that provides the attraction to the electrical energy of the Sun that keeps both in sync and harmony in the Aether.

Tesla proved electromagnetism to exist everywhere in all of Nature. He, along with the work of unsung heroes like Charles-Augustin de Coulomb who proved the existence of the Aether everywhere in Nature. Later, J.J. Thompson showed how electromagnetic energy created self-perpetuating energy and Nicolaus Tesla developed a working model with his Tesla Shield proving everything is made up of frequency, vibration and energy. Viktor Schauberger proved that Nature creates energy by implosion, not explosion, as can be seen in lightning strikes where energy is concentrated and focused to a single point.

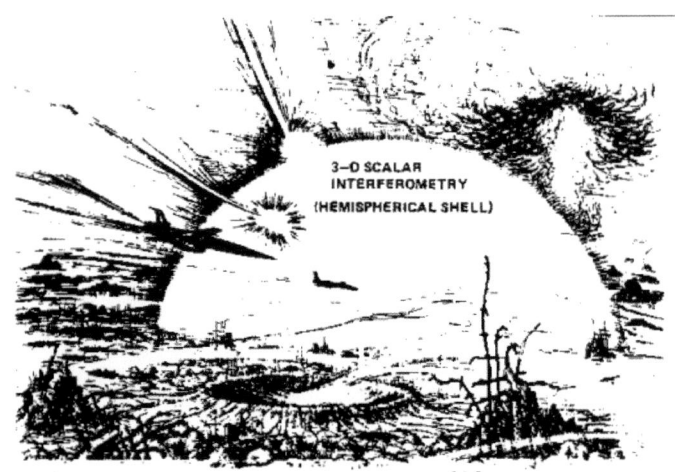

The Tesla Dome

The Saryshagan howitzer actually is a huge Tesla scalar interferometer with four modes of operation. One continuous mode is the Tesla shield, which places a thin, impenetrable hemispherical shell of energy over a large defended area. The 3-dimensional shell is created by interfering two Fourier-expansion, 3-dimensional scalar hemispherical patterns in space so they pair-couple into a dome-like shell of intense, ordinary electromagnetic energy. The air molecules and atoms in the shell are totally ionized and thus highly excited, giving off intense, glowing light. Anything physical which hits the shell receives an enormous discharge of electrical energy and is instantly vaporized — it goes pfft! like a bug hitting one of the electrical bug killers now so much in vogue.

If several of these hemispherical shells are concentrically stacked, even the gamma radiation and EMP from a high altitude nuclear explosion above the stack cannot penetrate all the shells due to repetitive absorption and re-radiation, and scattering in the layered plasmas.

In the continuous shield mode, the Tesla interferometer is fed by a bank of Moray free energy generators, so that enormous energy is available in the shield. Electromagnetic radiation is energy that is propagated through free space or through a material medium in the form of electromagnetic waves, such as radio waves, visible light, and gamma rays. The term also refers to the emission and transmission of such radiant energy.

The wavelength of the light determines its characteristics. For example, short wavelengths are high energy gamma-rays and x-rays, long wavelengths are radio waves. The whole range of wavelengths is called the electromagnetic spectrum.

Domed Earth is Like a Giant Electromagnetic Speaker

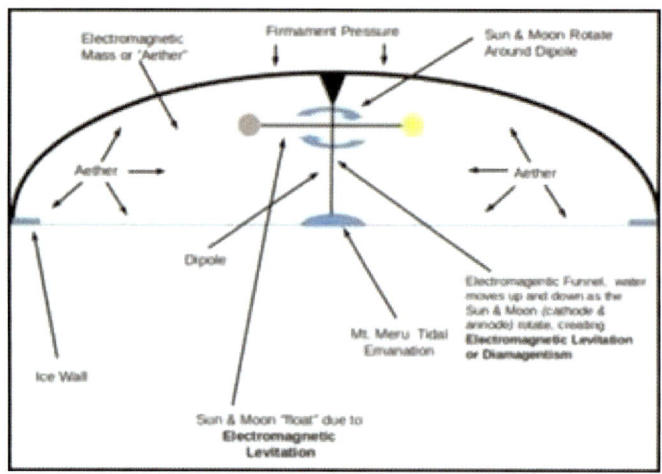

Ring magnets, like the kind found in common loudspeakers, have a central North pole with the opposite 'South' pole. The Flat Earth model projects similar electromagnetic properties in that the Arctic Circle provides the outer ring magnet and the North Pole, known in mythology as Mt. Meru, is the central opposite charge. This perfectly demonstrates the properties of how energy is distributed on a Flat Earth.

Round ball Earth theory states our alleged source of magnetism is emitted from a hypothetical, unproven molten magnetic core in the center of the ball which physicists claim conveniently cause both poles to constantly move, thus evading independent verification at their two 'ceremonial poles.'

The human energy field also reside within Tesla's electromagnetic fields and connect us to the trinary (Earth, Sun and, Moon). The Sun acts as the positive (+) electromagnetic energy causing 'low-tide' and the moon acting as the negative (-) electromagnetic energy. This alternating current also causes the high and low tides of the oceans. The reason that lakes, rivers and streams do no experience tidal action is due to salinity, or salt. The ocean conducts the needed sodium to connect the negative charge magnetic (anode) of the North Pole with the positive charge (cathode) of the Antarctic region, just like how a common ring magnet loudspeaker work.

Newtonian gravity theory claims that size matters relative to water. This is only reason given for why the oceans are effected by lunar gravitational pull but the largest lakes and rivers are not. Gravity is unseen and unproven until very recent findings of so called "gravitons." "*Graviton* is a hypothetical elementary particle that mediates the force of gravitation in the framework of quantum field theory". ~ Wiki

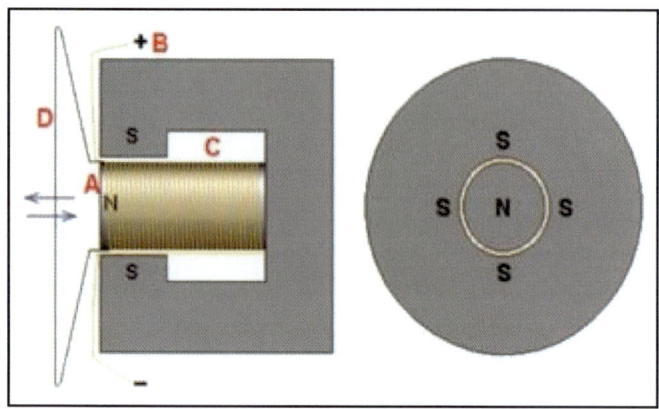
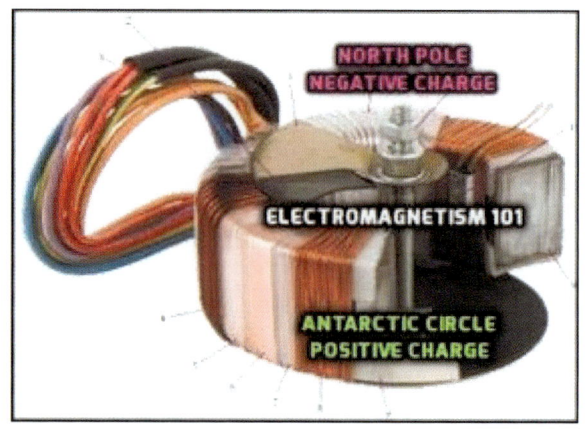

There is Only One Magnetic (North) Pole

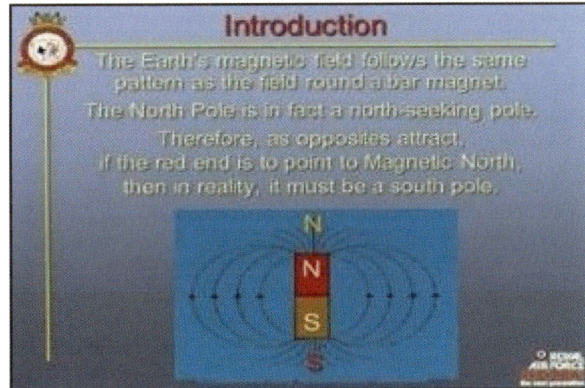

A compass only will point North, not East, West or South. It is not even true North due to the chaotic nature of the magnetic interaction inside the Earth's core between the North and South Poles. The modern mariner's compass is an impossible and useless instrument for use on a round ball-Earth. It simultaneously points North and South over a flat surface, yet claims to be pin-pointing two constantly moving geomagnetic poles at opposite ends of a spinning sphere originating from a hypothetical molten metal core. If compass needles were drawn to the North pole of a globe, the opposing 'South' needle would be pointing straight up to the heavens, yet never does.

If one were to stand with a compass at the North pole, then we would see the compass needle spin in a complete circle, yet no one has ever shown this to occur. Science 'assumes' (there is that word again) that there is a bar magnet buried deep inside Earth and due to this assumption, the core of the Earth 'must' be made of a nickel/iron core for the magnet to work properly. The North central pole is the only proven fixed point on our flat Earth. The South being all straight lines outwards from the pole, East and West being concentric circles at constant right angles 90 degrees from the pole. If one were to westerly circumnavigate Earth, the Polaris star would continually be on your right, while an easterly circumnavigation is going around with the Polaris star always at your left.

On a Flat Earth model, the compass will point towards the center of the circular flat earth only with no anomalies or corrections for "true North" readings. ~ Eric Dubay, Flat Earth Conspiracy

The Great Antarctic Ice Walls

On Flat Earth models, the outer edge of the Earth is bounded all around by an Ice Wall said to be anywhere from 150 to 300 feet in height. This wall prevents the oceans from spilling over the side of the Earth. The exact size of the Ice Wall varies between different Flat Earth Models. The Antarctic mountains soar up on 10,000 feet of all snow and ice, according to Freemason Admiral Byrd.

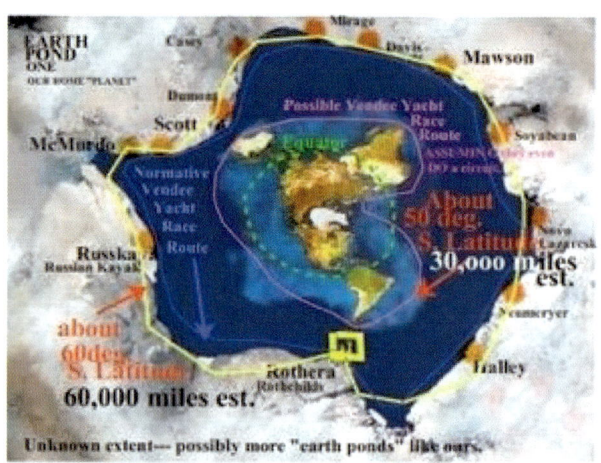

This view of the Ice Wall is generally agreed to correspond to the coastline of Antarctica in the round ball model of Earth. It is unknown whether the ice extends outward forever or there is an actual boundary to the edge of the electromagnetic dome that meets at Earth level.

The Antarctic is the coldest place on Earth yet it is closest to, and surrounded by, the warmest places on Earth; South America, South Africa, Australia and New Zealand. This makes no sense whatsoever if Antarctica is closest to the Sun for six months out of the year in a heliocentric model. Artist ~ Rick Potvin

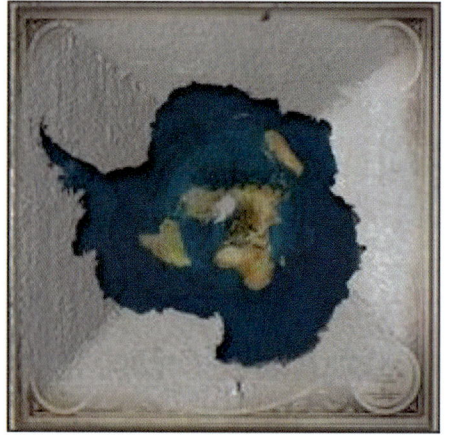

The Luminaries Behind the Dome
Twinkle, twinkle little star, How I wonder what you are?
Unretouched Actual Photographs

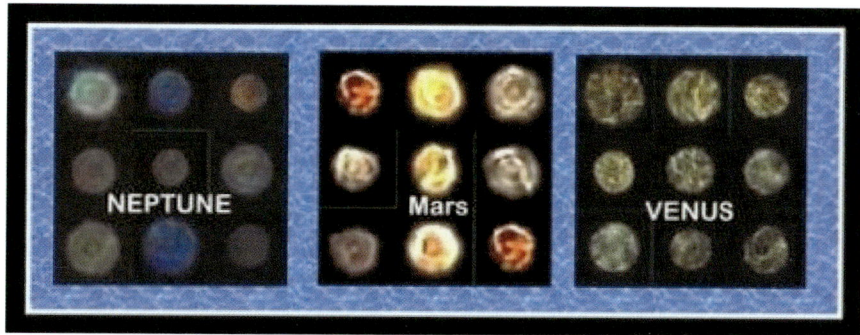

Finding the occasional straw of truth awash in a great ocean of confusion and bamboozle requires intelligence, vigilance, dedication and courage. But if we don't practice these tough habits of thought, we cannot hope to solve the truly serious problems that face us – and we risk becoming a nation of suckers, up for grabs by the next charlatan who comes along. One of the saddest lessons of history is this: **If we've been bamboozled long enough, we tend to reject any evidence of the bamboozle.** *We're no longer interested in finding out the truth. The bamboozle has captured us. It is simply too painful to acknowledge—even to ourselves— that we've been so credulous.* —Carl Sagan

For less than $1,000, you can buy a Nikon Coolpix P900 camera and take many of the same photographs of the stars behind our watery dome firmament as you see in the unretouched photographs above. You can view the colors of Mars, Jupiter, Venus and many of the brightest stars in the sky when the camera is in full zoom mode. The stars are not physical but light energy beings as you can view for yourself. This is a simple, do-it-yourself empirical proof that the heliocentric round ball theory is a fabricated lie. According to ancient mythology and legend, the stars were called our *'Luminaires'* who have ascended the physical realm and are in set up in the heavens to help guide and direct us down here on Earth for our physical, mental and spiritual development.

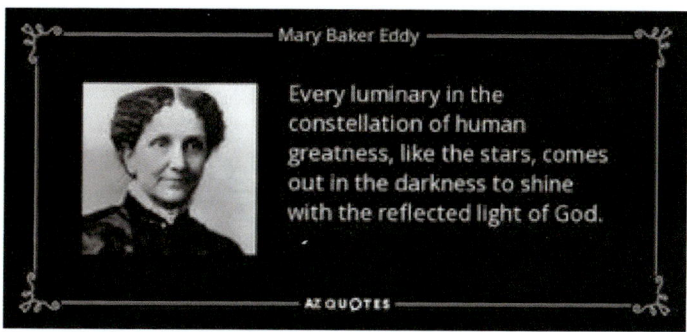

A Rapidly Spinning Globe Earth Ball?

According to heliocentric theory the Earth is rotating on its axis at 1,000 miles per hour. Every hour of every day, yet we never feel a whisper of wind or movement underfoot.

Sometimes the simple illogic of it all is all you need to convince yourself that gravity is a complete and utter hoax on all modern humanity. Not only are we told by modern science about such incredible speeds of rotation but what makes us rotate so fast is said to be perpetual motion left over from our expulsion from a cavity in the Sun 4.5 billion years ago, when Earth spit out the Moon from her womb.

For Earth to orbit one time around the Sun in one year means we would have to travel some 533 million miles each and every year to get back to the same spot for our birthdays. This calculates out to even more incredible speed the Earth must be moving around the Sun of over *1,000 miles per second*, yet we still never feel a thing.

"In short, the sun, moon, and stars are actually doing precisely what everyone throughout all history has seen them do. We do not believe what our eyes tell us because we have been taught a counterfeit system which demands that we believe what has never been confirmed by observation or experiment.

That counterfeit system demands that the Earth rotate on an 'axis' every 24 hours at a speed of over 1000 MPH at the equator. No one has ever, ever, ever seen or felt such movement (nor seen or felt the 67,000MPH speed of the Earth's alleged orbit around the sun or its 500,000 MPH alleged speed around a galaxy or its retreat from an alleged 'Big Bang' at over 670,000,000 MPH!). Remember, no experiment has ever shown the earth to be moving. ~ Marshall Hall

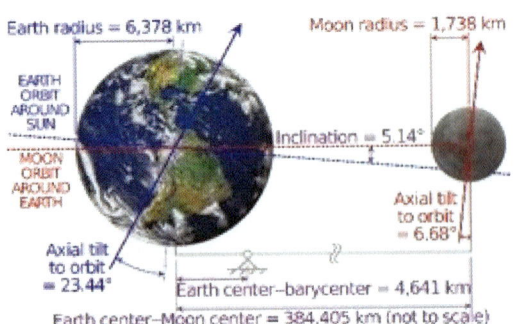

If Earth is tilted on its back at 23.5 degrees, wobbling on its axis, this means gravity must be increasing on the backside of the tilt to hold the might oceans in while we are rotating at 1,000 miles per hour. Exactly what mechanism of gravity activates to increase gravitational pull when needed has never been proven either. Think of the trillions of gallons of water that would have to be held in by invisible, unprovable, just believe the science, "gravity"!

"Yet, how can gravitation increase to hold at the backside of a tilted Earth of 23.5 degrees, yet not even be measured on the largest lakes in the world? The moon has 1/6 the gravitational pull of the Earth, yet it moves our oceans up and down, twice each day and actually changes the shape of a mass 4X greater than itself from 238,000 miles away. This is achieved despite the greater Earth's pull and the even greater gravitational pull by the Sun some 93million miles away that locks in both the Earth and moon in his orbit."—Eric Dubay

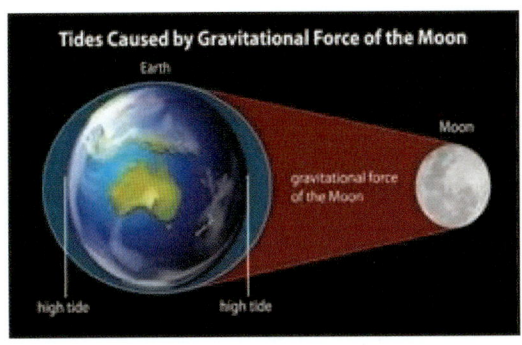

All alleged planets are said to be 'falling' in the vacuum of space, yet how is it that the smaller mass doesn't fall all the way into the larger mass if all is falling in attraction towards each other? What is keeping the Moon locked in consistent orbit with Earth if it is in a state of 'falling'? How can something pulling prevent something from falling at the same time?

"Indeed, tides exist in all bodies of water, even one's bathtub, but is so infinitesimally small, as to be unmeasurable. Even on Lake Superior, the largest of the Great Lakes of North America, the tiny effect of a tide is overcome by the effect of barometric pressure and the phenomenon known as a seiche. There are no Tide Tables of the Great Lakes and seiche warnings are rarely broadcast, as most cause a variance of less than 50 cm. The effects of a seiche may be felt strongest in the Straits of Mackinac between Lakes Huron and Michigan." —R. Manning, Marquette, NASA Astrophysicist

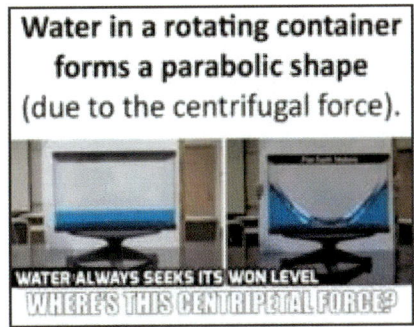

"How is it that a law of gravitation can pull up a toy balloon and cannot put up a brick? I throw up this book. Why doesn't it go on up? That book went up as far as the force behind it forced it and it fell because it was heavier than the air and that is the only reason. I cut the string of a toy balloon. It rises, gets to a certain height and then it begins to settle. I take this brick and a feather. I blow the feather. Yonder it goes. Finally, it begins to settle and comes down. This brick goes up as far as the force forces it and then it comes down because it is heavier than the air. That is all."—Wilbur Voliva

Gravity Doesn't Exist; It's Simply Buoyancy, Weight & Density

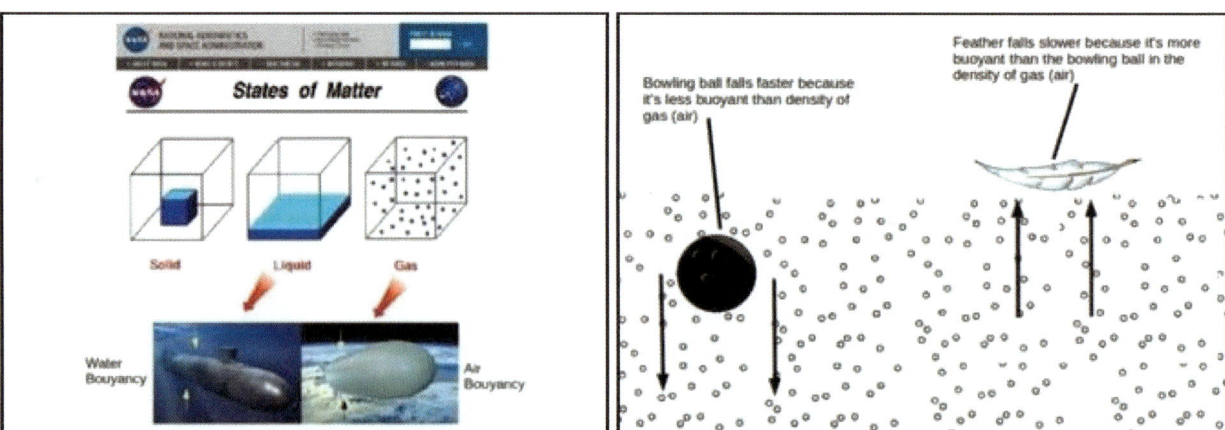

"For this reason, an object whose density is greater than that of the fluid in which it is submerged tends to sink. If the object is either less dense than the liquid or is shaped appropriately (as in a boat), the force can keep the object afloat." ~ Wikipedia

If you speak to Astrophysicists they will tell you about tiny, minute, sub atomic particles called "gravitons" that have only been discovered in the second decade of the 21st century and are said to be what creates gravity. Again, this is all coming from technology that is said to be so highly sophisticated that few can understand how this is done yet is cited as new evidence that gravity does exist. Gravity cannot be seen, nor proven in lab tests except in a vacuum yet Earth is not a vacuum.

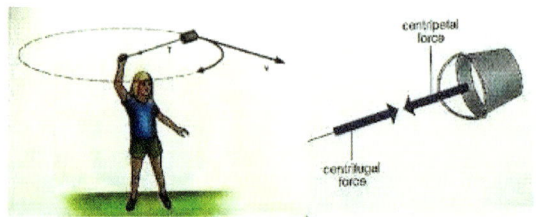

Most in school are taught about gravity from an example of a person in the center swinging a rope with a bucket of water attached to show how water is held in by gravity. This allegedly shows us how gravity works on Earth yet Earth is not holding a rope, nor is it a pail with an outer metal enclosure so how can this be proof of gravity?

Simply, the weight, the size and how dense and buoyant something is, determines if something falls or elevates. Objects will seek its own equilibrium and balance as in Nature. Nature balances, that's what she does. A rock in a pond will sink while a beach ball will float. Density, weight and buoyancy.

The only difference between water (liquid) and air (gas) is the density. Buoyancy is not confined to just water; it also pertains to "Air buoyancy" or the buoyancy of air. Similar to objects at the bottom of the ocean of water looking upward at objects floating above it, humans live at the bottom of an "ocean" of air and look upward at balloons drifting above us. Objects in water are buoyed up because the pressure acting up against the bottom of the object 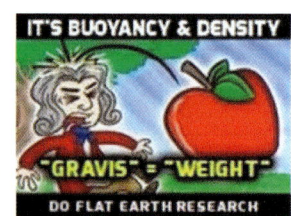 exceeds the pressure acting down against the top. Likewise, air pressure acting up against an object in air is greater than the pressure above pushing down. The buoyancy, in both cases, is equal to the weight of fluid displaced – Archimedes' principle holds for air just as it does for water.

94

What is the Source of Gravity?

"What is gravity?" asked the interviewer.

"I have no idea," Neil DeGrasse replied. "Okay, next question."

Then, he went on to explain, "Here's the difference. We can describe gravity, we can say what it does to other things."

"We can measure it, we can predict with it," he continued, "but when you start asking, like, what it is? I don't know."

His cohost Leighann Lord said, "So I accidentally asked a deeper question than I meant to?"

"No, no," replied Tyson, "you were meant to ask deep questions in life. So…in an Einstein-ian answer, we'd say gravity is the curvature of space and time, and that objects will follow the curvature of space-time and we interpret that as a force of gravity. That's probably the best answer I can give to a 'What is gravity?' question." —Astrophysicist Neil de Grasse Tyson on his Star Trek podcast and radio show, 6.12.14

> *"Gravity is completely different from the other forces described by the standard model. When you do some calculations about small gravitational interactions, you get stupid answers. The math simply doesn't work."* —Mark Jackson, Theoretical Physicist

The source of gravity is defined by modern science as: **"the mutual attraction principle determined solely by density, mass."** The word weight is only used to measure the gravitational pull between two objects.

According to etymology, which defines the origin and root of words, 'gravis' means 'weight' from the Latin word gravitas, which means "weight" or "heaviness". Sir Isaac Newton's Law of Universal Gravitation completely omit the word "weight" from the accepted definition of gravity altogether!

> "Furthermore, this magnetic-like attraction of massive objects gravity is purported to have can be found nowhere in the natural world. There is no example in nature of a massive sphere or any other shaped-object which by virtue of its mass alone causes smaller objects to stick to or orbit around it!

There is nothing on Earth massive enough that it can be shown to cause even a dust-bunny to stick to or orbit around it! Try spinning a wet tennis ball or any other spherical object with smaller things placed on its surface and you will find that everything falls or flies off, and nothing sticks to or orbits it. To claim the existence of a physical 'law' without a single practical evidential example is hearsay, not science." —Eric Dubay, Flat Earth Conspiracy

If the moon is 2,160 miles in diameter and the Earth 8,000 miles in circumference at the Equator, then using modern science's own math and laws it follows that the Earth is 87 times more massive. Therefore, the larger object should attract the smaller to it, and not the other way around! If the Earth's greater gravity is what keeps the Moon in orbit, it is impossible for the Moon's lesser gravity to supersede the Earth's gravity. If the Moon's gravity truly did overpower the pull of gravity from Earth's causing the tides to be drawn towards it, there should be nothing to stop them from continuing onwards and upwards towards their great attractor. Science calls this action the 'tidal bulge' creating a pear-shaped Earth, which previous was an oblate shape Earth before it was a round ball Earth. Astrophysicists simply cannot make up their minds on what the shape of Earth is…yet if we had a live camera on the Moon, we could see for ourselves!

Heliocentrism tells us that the Moon is tidal locked step with the Earth due to Earth's 4X larger size than its 6X greater gravitational pull than the Moon where the Earth's gravity keeps the Moon in an elliptical orbit around Earth.

The Karmin line in space is said to begin at 62 miles above Earth and where Earth's atmosphere meets outer space. This is where gravity is said to have lessened enough from Earth's pull that astronauts can space walk. We are told that the Moon's gravity, with 1/6th the gravitational force of Earth, travels 238,000 miles to move our oceans up and down twice a day.

The lunar-caused ocean tides rise greater in some areas and less in others while the round ball spins at 1,000 mph on a 23.5 degree tilt holding in trillions of gallons of water without ever releasing water outward, where gravity is lessened the further away from center one goes and it is all due to invisible, unproven gravity.

Newton's Laws of Motion

1. (*The Law of Inertia*) A body at rest remains at rest and a body in motion remains in motion with a constant speed and in a straight line, unless acted upon by an outside force.

2. The acceleration of an object is proportional to the force acting upon it and is directed in the direction of the force. That is, $F=ma$.

3. To every action there is an equal and opposite reaction.

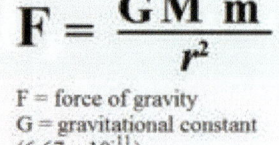

$$F = \frac{GMm}{r^2}$$

F = force of gravity
G = gravitational constant (6.67 x 10^{-11})
M = mass of one object
m = mass of other object
r = distance between the two objects

The velocity and path of the Moon are uniform and should therefore exert a uniform influence on the Earth's tides yet Earth's tidal rises and falls vary greatly around the world. Humans and other life on Earth are said to be held affixed to the spinning orb due to gravity yet we can escape gravity by simply jumping up in the air whenever we want.

The Moon's pull is so consistent that we can set our oceans tide tables almost up to the second, weeks and months in advance. Some areas have tidal changes of dozens of feet while only a short distance away, tidal change is as little as foot or two. Curiously, the Moon's gravity cannot be measured on all lakes, rivers, ponds and streams. The reason given is that size matters. We are never told what volume of size kicks in the magical gravitational forces on water just that it is somewhere between the mass of Lake Erie and the size of our Oceans.

Here's the "size matters" explanation from NASA, the sole source of all things space related. Note they have never proved gravity except in a vacuum. Once again, Earth is not a vacuum.

"Indeed, tides exist in all bodies of water, even one's bathtub, but is so infinitesimally small, as to be unmeasurable. Even on Lake Superior, the largest of the Great Lakes of North America, the tiny effect of a tide is overcome by the effect of barometric pressure and the phenomenon known as a seiche.

There are no Tide Tables of the Great Lakes and seiche warnings are rarely broadcast, as most cause a variance of less than 50 cm. The effects of a seiche may be felt strongest in the Straits of Mackinac between Lakes Huron and Michigan." —R. Manning, Marquette, NASA Astrophycisist

There Is No 'Vacuum of Space'

According to Wikipedia citing NASA: "There is no clear boundary between Earth's atmosphere and space, as the density of the atmosphere gradually decreases as the altitude increases. There are several standard boundary designations, namely: The Federation Aeronautique Internationale has established the Karmin line at an altitude of 100km (62mi) as a working definition for the boundary between aeronautics and astronautics. This is used because at an altitude of about 100km (62 mi), as Theodore Von Karmin calculated, a vehicle would have to travel faster than orbital velocity in order to derive sufficient aerodynamic lift from the atmosphere to support itself. The United States designates people who travel above an altitude of 50 miles (80 km) as astronauts."

Due to Earth's atmosphere, we are told there exists friction and drag that allows propulsion. A car pushes off the road as the wheel's spin, a plane pushes against the stratosphere, a boat against water, etc.... so what does a space craft push off of in the 'vacuum of space if it is void of drag and friction? Vacuum is defined as space void of matter.

An approximation to such vacuum is a region with a gaseous pressure less than atmospheric pressure. Physicists often discuss ideal test results that would occur in a perfect vacuum, which they sometimes simply call 'vacuum' or free space, and use the term 'partial vacuum' to refer to an actual imperfect vacuum as one might have in a laboratory or in space."

The word vacuum stems from the Latin adjective "vacuus" for 'vacant' or 'void.'

A vacuum in the lab is created by sucking out the gases/matter, but in space the same vacuum is supposedly created and maintained by the paradoxical expansion of the universe. Imagine exhaling into a balloon (expansion) and then inhaling the air back out (vacuum) they are two different processes yet NASA says it yields the same results! The only proof of the vacuum of space is the illusion of weightlessness, and without the illusion of weightlessness in space, it would reveal the farce and canard that NASA calls the 'Vacuum of Space.' One cannot have a vacuum next to a non-vacuum and have interplay between the two, it's impossible and paradoxical.

Atmospheric pressure on Earth is said to be around 14.7 pounds-per-square inch or 14psi. A vacuum of space has zero psi. How do we magically go from 14psi to 0 psi is unexplained by physicists?

Heliocentrics Analemma Paradox

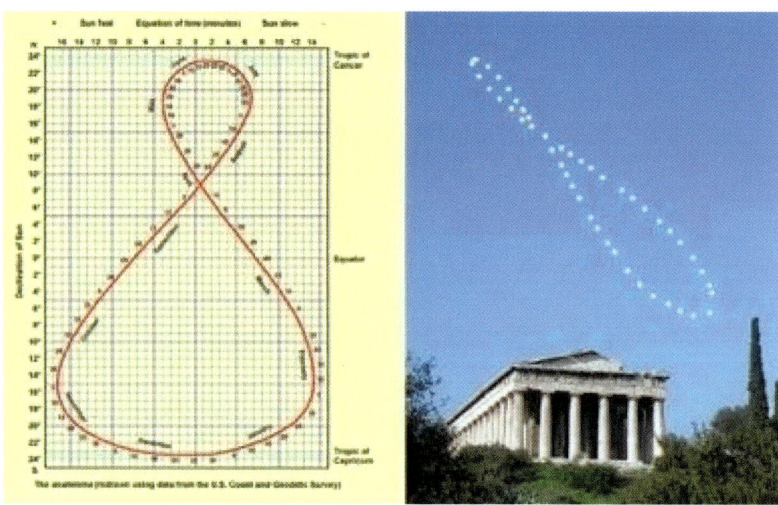

*"In astronomy, an analemma, Greek for "support," is a diagram showing the deviation of the Sun from its mean motion in the sky, **as viewed from a fixed location on the Earth.**"* ~Wikipedia

The Solar Analemma proves the Flat Earth on the Azimuthal Equidistant (AE) map. All points on the map are at proportionately correct distances from the center point, and all points on the AE map are at the correct azimuth (direction) from the center point.

If you photographed the sun's position in the sky at the same time every day for a year, you would observe a figure 8 shape called the 'Analemma.' The flat earth model explains this phenomenon perfectly. The sun circulates around flat earth along the Tropic of Cancer in the summer and then moves south to the Tropic of Capricorn in the winter months.

The Sun also circles the earth along a much wider path in the winter as it is moved farther south. You can see this in the chart above because the Analemma is wider in the lower part of the figure eight. The Sun circles above the Earth closer to the North Pole in the summer and therefore has a much tighter circulation. This is represented by the smaller section of the figure eight at the top of the Analemma.

This cannot be explained on the globe. They say the figure eight is created by the earth's tilt. But why is the Analemma wider on the bottom and smaller on top if Earth is a round nearly uniform ball?.

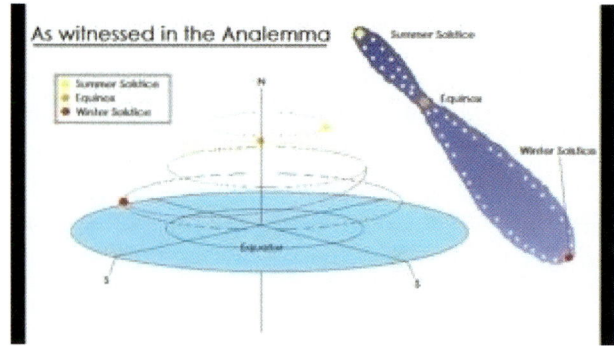

Stars Rotate Above, Earth is Still

"It is found by observation that the stars come to the meridian about four minutes earlier every twenty-four hours than the sun, taking the solar time as the standard. This makes 120 minutes every thirty days and twenty-four hours in the year. Hence all the constellations have passed before or in advance of the sun in that time. This is the simple fact as observed in nature, but the theory of rotundity and motion on axes and in an orbit, has no place for it. Visible truth must be ignored, because this theory stands in the way, and prevents its votaries from understanding it."
— *"Earth Not a Globe!"* by Samuel Rowbotham

—

"NASA and modern astronomy say Polaris, the North Pole star, is somewhere between 323–434 light years away from Earth or about 2 quadrillion miles. Firstly, note that the calculations of scientific proof vary by over six hundred trillion miles! If modern astronomy cannot even agree on the distance to stars within hundreds of trillions of miles, perhaps their 'science' is flawed and their theory needs re-examining. However, even granting them their obscurely distant stars, it is impossible for helio-centrists to explain how Polaris manages to always remain perfectly aligned straight above the North Pole throughout Earth's various alleged tilting, wobbling, rotating and revolving motions while hurtling at incredible speeds around the Sun and Milky Way.

If Earth were a spinning ball it would be impossible to photograph star-trail time-lapses turning perfect circles around Polaris anywhere but the North Pole. At all other vantage points the stars would be seen to travel more or less horizontally across the observer's horizon due to the alleged 1,000 mph motion beneath their feet. In reality, however, Polaris's surrounding stars can always be photographed turning perfect circles around the central star all the way down to the Tropic of Capricorn".
—Eric Dubay, *Flat Earth Conspiracy*

Observation Experiments We Can All Do

Much of Flat Earth proof is available to those that wish to confirm FE theory for themselves using their own observation, thought, reason, logic and common sense. Here are some examples all can do for themselves.

1. During the 1st quarter waxing Moon, when you can see it rise over the Eastern horizon, observe the Sun at sunset relative to the light shining directly atop the moon. The Sun is to your opposite shoulder than the Moon over your left shoulder. First, the light on the Moon doesn't change relative to the movement away from Earth and Moon. Second, the light of the Moon remains coming from the top even though the Sun is lower on the western horizon. Shadows move as the Sun moves on Earth yet never moves on the Moon? At Sun's setting, you will clearly see the light is coming from directly above and impossible to be lit by the setting Sun down and to your right.

2. Take a plastic bowl and place face down into a sink over the drain hole. Hold down and turn on the water until the entire bowl is covered with water and no water is draining out. Once water has filled to the top of the bowl you can let go and see how Earth could be inside a dome filled with air, yet completely under water at the same time.

3. Look at the moon during its half-phase through a telescope, or high zoom camera, and see stars for yourself through the dark portions of the moon.

4. Go to the beach or a mountain top with unobstructed views. On a clear day at Sea-*level* you can see some 10 miles scanning from side to side and about 3 miles straight ahead. Basic spherical geometry says we should see at least 66 feet of curvature with a cresting ocean front and center of your viewing. Go up to mountain top and estimate how far you can see. Multiply miles x miles x 8 inches to get the curvature. Do you see a curve anywhere?

5. As you increase in altitude in an aircraft or hot air balloon does the horizon fall away from your vision? The Earth never falls away from your eye yet the ground rises to meet your eye's viewpoint proving Earth not a globe but a flat plane…as observed from the aero- plane!

6. Observe clouds on sunny days and clouds on full moon days. Sometimes you will see clouds behind the Sun and the Moon. How can this be if the Sun is said to be 93 million miles away and the Moon 238,000 miles away?

Take a high-powered camera or binoculars or telescope to a large lake or ocean. Watch a ship sail away from you over the horizon with your eyes. Then take your magnifier and bring the ship back into view, proving the Earth is not a round ball. Lay out at night and stare at the Northern Star. Watch all the other stars circle around the fixed North Star. Except for the fixed North Star, all stars move around us on Earth.

Flat Earth Science Hall of Heroes

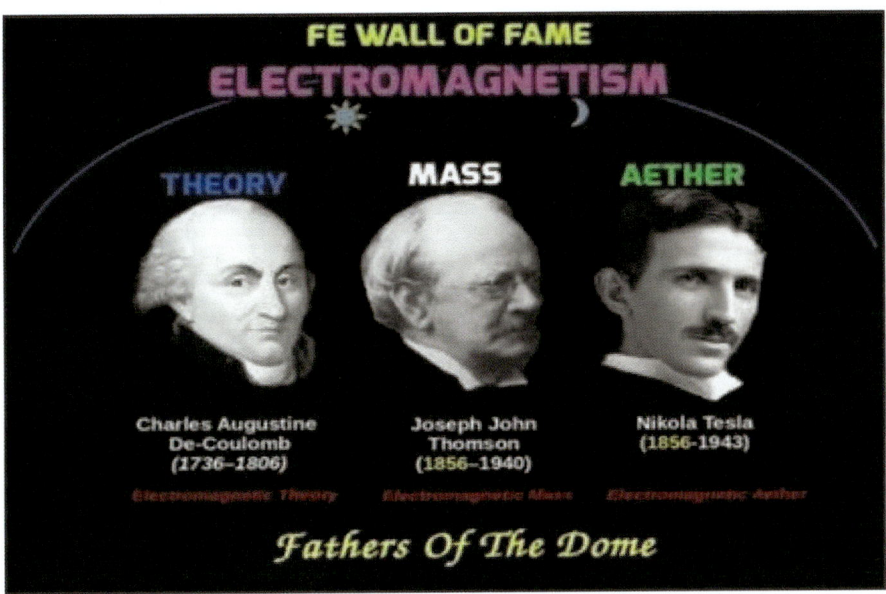

Each of these scholarly scientists proved the Earth to be made up of Electromagnetism and should be celebrated heroes rather than Copernicus, Newton and Einstein, who were all puppets to the Jesuits and the Roman Catholic Church who created the heliocentric theory.

Charles-Augustin de Coulomb; Father of Electromagnetic Discovery. (1736 –1806) Charles-Augustin de Coulomb was an eminent French physicist. He formulated Coulomb's Law which showed the electrostatic interaction between electrically charged particles proving electromagnetism exists in the Aether. A Coulomb, is a unit of electric charge, named after him. He also developed the inverse square law of attraction and repulsion of unlike and like magnetic poles. This laid out the foundation for the mathematical theory of magnetic forces formulated by French mathematician Siméon-Denis Poisson.

John J. Thomson; Father of Electromagnetic Self-perpetuating Energy (1856 –1940) Mr. Thomson was awarded the 1906 Nobel Prize in Physics for the discovery of the electron and for his work on the conduction of electricity in gases. Eight of his students, and his son George Paget Thomson, became Nobel Prize winners, either in physics or in chemistry. Electromagnetic mass was initially a concept of classical mechanics, denoting as to how much the electromagnetic field, or the self-energy, is contributing to the mass of charged particles. In 1881 Mr. Thomson showed showed the dynamic interplay of inertial mass induction. Due to this self-induction effect, electrostatic energy behaves as having some sort of momentum and 'apparent' electromagnetic mass, which can increase the ordinary mechanical mass of the bodies. This increase arises from the previously unknown electromagnetic self-energy.

Nikola Tesla; Father of Free Electromagnetic Energy for All (1856 –1943)
"Tesla's discovery can eventually remove every conceivable external human limitation. If we humans ourselves can elevate our consciousness to properly utilize the Tesla electromagnetics, then Nikola Tesla—who gave us the electrical twentieth century in the first place—may yet give us a fantastic new future more shining and glorious than all the great scientists and sages have imagined."—Tom Bearden

Chapter 4

Flat Earther's Claims of a Cons-Piracy

**Just one live camera of Earth from the Moon
ends forever the entire Flat Earth discussion!
Can you hear us NASA?**

Computer Generated Imagery (CGI)
Or How the NASA Military Had Photoshop Long Before Adobe

Mr. Rob Simmons, an employee of NASA, said in a statement in April of 2015 the following: "**My part was integrating the surface, clouds, and oceans to match people's expectations of how Earth looks from space. That ball became the famous Blue Marble.**"

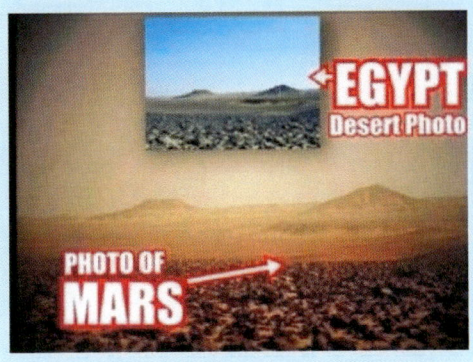
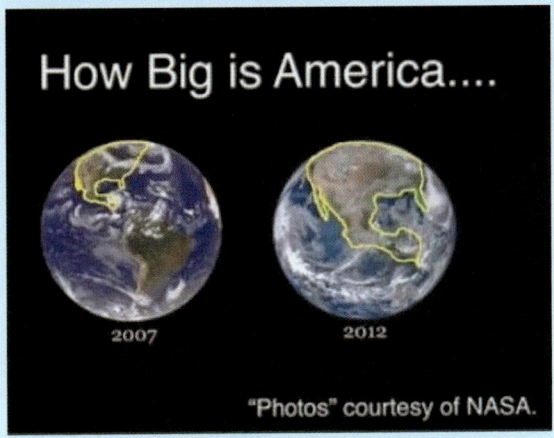

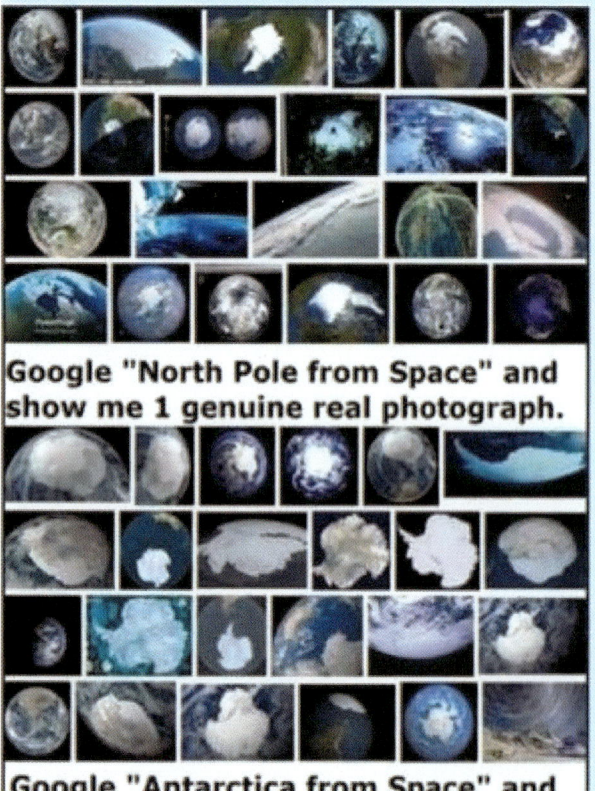

Why Doesn't the Blue of Earth Reflect onto the Moon Surface?
How Can Golden Warm Sunlight Reflect Off A the Dusty Gray Surface of the Moon?

The Sun, from 93 million miles away, is said to reflect off the Moon onto Earth. The Moon is 238,000 miles from Earth and from a gray dusty surface then reflects yellow/white moonlight evenly across the round ball Earth.

The big question is then, why doesn't the deep "Blue Marble" of Earth shine blue onto the Moon if it is 4X larger than the Moon and also reflecting off of the Sun nearly the same distance away? I have never heard anyone ask this most basic question before, much less get a plausible answer.

Gray is one of the least reflective surfaces after black. Science tells us that the pale gray round ball Moon reflects Golden Sun evenly across the surface of Earth on a full Moon. The full Moon light is so bright that one can see clearly at night. The light is silvery and cool, not warm and golden.

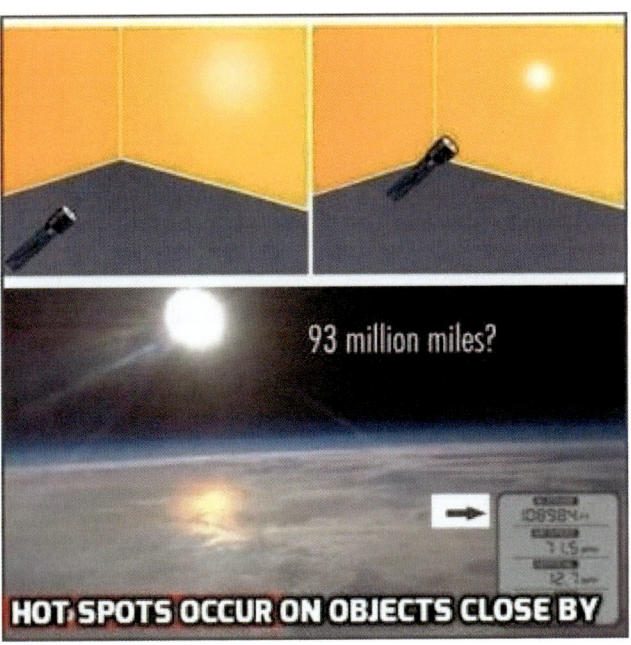

1) How can a round ball reflecting the Sun not have a hot spot?

2) How can something round reflect evenly across another round ball object?

3) How could they walk on the moon, if the Sun reflection was so intense that it can light up the Earth at night from 238,000 miles away at night?

4) If the Earth is 4x larger than the Moon, why doesn't the "Blue Marble" reflect blue onto the Moon?

Why is Moonlight Warmer in the Shade Than in Direct light?

 For around $30 US dollars you can go to your favorite hardware store and pick up a thermo-laser that will measure the temperature of whatever you aim it at.

During a full Moon you can take heat readings of both the full moonlight and moon shadows. You will discover that the Moonlight in the shadows are several degrees warmer than in the direct sunlight. This simple experiment proves that the Sun cannot be reflecting off the Moon as we have been told.

"The Sun's light is golden, warm, drying, preservative and antiseptic, while the Moon's light is silver, cool, damp, putrefying and septic. The Sun's rays decrease the combustion of a bonfire, while the Moon's rays increase combustion. Plant and animal substances exposed to sunlight quickly dry, shrink, coagulate, and lose the tendency to decompose and putrify; grapes and other fruits become solid, partially candied and preserved like raisins, dates, and prunes; animal flesh coagulates, loses its volatile gaseous constituents, becomes firm, dry, and slow to decay. When exposed to moonlight, however, plant and animal substances tend to show symptoms of putrefaction and decay.

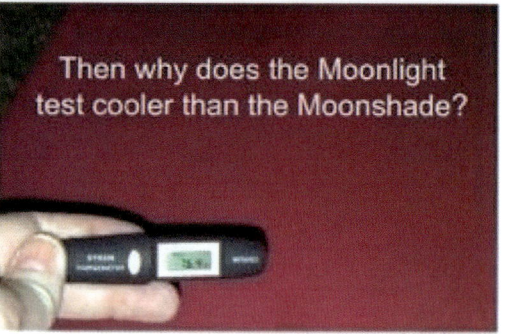

In direct sunlight, a thermometer will read higher than another thermometer placed in the shade, but in full, direct moonlight a thermometer will read lower than another placed in the shade. If the Sun's light is collected in a large lens and thrown to a focus point it can create significant heat, while the Moon's light collected similarly creates no heat. In the "Lancet Medical Journal," from March 14th, 1856, particulars are given of several experiments which proved the Moon's rays when concentrated can actually reduce the temperature upon a thermometer more than eight degrees.

So, sunlight and moonlight clearly have altogether different properties, and furthermore the Moon itself cannot physically be both a spherical body and a reflector of the Sun's light! Reflectors must be flat or concave for light rays to have any angle of incidence; If a reflector's surface is convex then every ray of light points in a direct line with the radius perpendicular to the surface resulting in no reflection."

~ Eric Dubay

Satellites Perform Perfectly in + 3,000 F Temperatures???

According to NASA, there are some 1,265 functioning satellites orbiting Earth to provide various functions such as military applications, global positioning satellites, cell phone communications, Earth tomography, weather predictions, asteroid/comet warnings, etc. The average satellite allegedly in space is made up of carbon fiber, aluminum-beryllium alloy graphite, titanium, Kevlar, glass, and sheet metal. Inside are batteries, computers, cameras and highly integrated electronic circuitry that

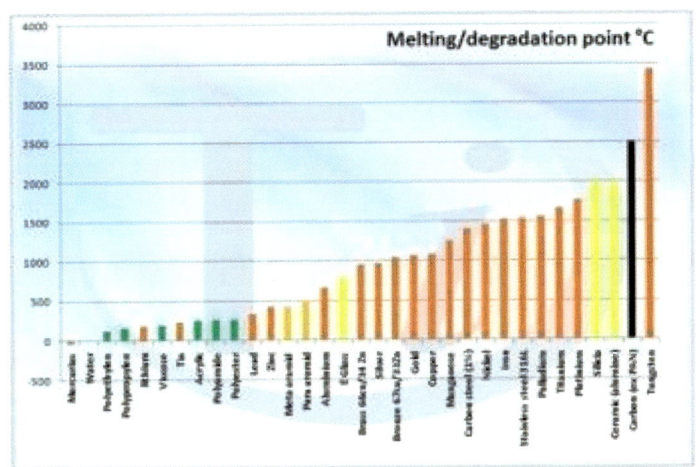

operate the satellite. Extended off the satellite are photo-voltaic solar panels that capture sunlight to keep continuous power that protrude and extend up to 50 feet connected by extension rods. Solar panels are made of glass, silicon, rare Earth materials, gold reflecting metal and special bonding cement.

These satellites are said to orbit the Earth some 22,500 miles up in space in the Thermosphere while traveling around Earth at a constant speed of 17,450 mph (23,000 ft./sec) to keep in geosynchronous orbit.

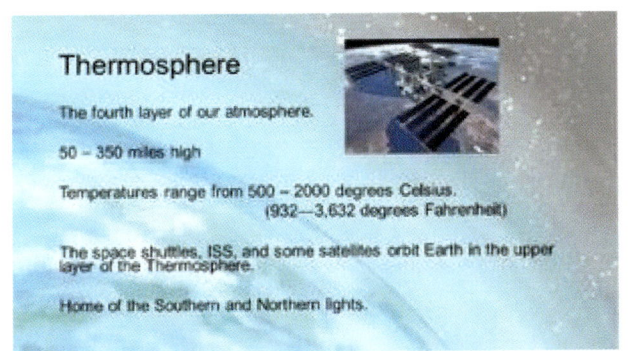

Temperatures in the Thermosphere reach over 3500 degrees Fahrenheit (2000 Celsius). The chart above shows the actual melting/degradation temperatures of different metals. The only elements in the periodic table that can withstand 2000°C are carbon, niobium, molybdenum, tantalum, tungsten, rhenium, and osmium. Except for carbon, these metals are very, very heavy and extremely conductive to heat. Most are very ductile when heat treated meaning they bend and coil. Carbon even has the highest thermal conductivities of all known materials! So, if you want to cook someone very efficiently and quickly, there is nothing better than a space capsule made out of graphite. Solar panels are fragile and made of many materials that only could withstand much lower temperatures than thousands of degrees Fahrenheit as well as be susceptible to space debris while their extended arms hurtle through space at such inconceivable speeds as 17,450 mph.

Dodging Deadly Micro-Bullets at 17,000 mph in Space?

*Note top middle photo: The picture of a space debris hole made through an aluminum block shot at 7 km/s (the orbital velocity of the ISS) made the 15 cm (5 7/8 in) crater in a solid block of aluminum).

It is impossible for satellites to exist in the Thermosphere of space due to extreme heat (over 3,000 F) and speeds of ridiculous travel (17,500 mph). The same applies to the alleged ISS space station. According to Wikipedia, there are millions of micro-meteors traveling at speeds as fast as 6,000 mph: "More than *500,000 pieces of space debris orbit Earth, traveling at speeds up to 175,000 mph*. A small piece of space debris traveling at these velocities could significantly damage a spacecraft or a satellite. It could also pose a threat to the lives of astronauts living on the International Space Station." At the low altitudes at which the ISS is said to orbit at altitudes of 205–270 miles (330–435 km) above Earth on average, there is a mass variety of space debris, consisting of a multitude of different objects including *entire spent rocket stages, defunct satellites, explosion fragments made up of anti-satellite weapon tests, paint flakes, slag from solid rocket motors, and coolant released by USA nuclear-powered satellites.* These objects, in addition to naturally occurring micro-meteoroids, space dust, and comet tailings are all said to be flying around Earth's orbit. NASA confirms this, yet not one report ever of any astronaut ever being hit, much less killed by any of the millions of space debris, nor even more than minor damage to any of the thousands of satellites said to be in continual orbit over the decades. (According to NASA, they have a tracking system that allows for '*Debris Avoidance Maneuvers*' called D.A.M., I kid you not!).

1. Why would NASA, or any other space agency allow the risk of death to their astronauts and space shuttles when just a sliver of paint particle could kill or cripple the spacecraft?
2. How is it that with over hundreds of walks in space, not one incident has ever been reported of space debris hitting an astronaut or disabling a mission while there are over half-of-a-million pieces in orbit?
3. From Wikipedia: "Space debris objects are tracked remotely from the ground, and the station crew can be notified. This allows for a DAM to be conducted, which uses thrusters on the Russian Orbital Segment to alter the station's orbital altitude, avoiding the debris. DAMs are not uncommon, taking place if computational models show the debris will approach within a certain threat distance."

How Can the Moon Rotate on its Axis?
Yet We Never See the Backside?

An unalterable rotational velocity thru all phases of planetary evolution is manifestly impossible. The truth is, the so-called "axial rotation" of the moon is a phenomenon deceptive alike to the eye and mind and devoid of physical meaning. The moon does rotate, not on its own, but about an axis passing thru the center of the earth, the true and only one." ~ Nikola Tesla

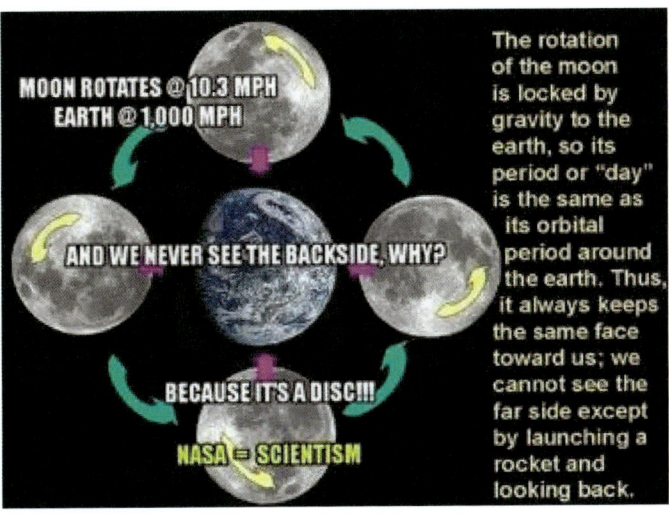

According to NASA, the Moon is rotating on its axis at 10.3 miles-per-hour. The Earth we're told is rotating at over 1,000 m.p.h. at the Equator. This means the Earth spins 9X to every on rotation of the Moon, yet we never see the backside of the Moon. How can this be?

High Noon Should Flip Every 6 Months
If We Are Orbiting the Sun

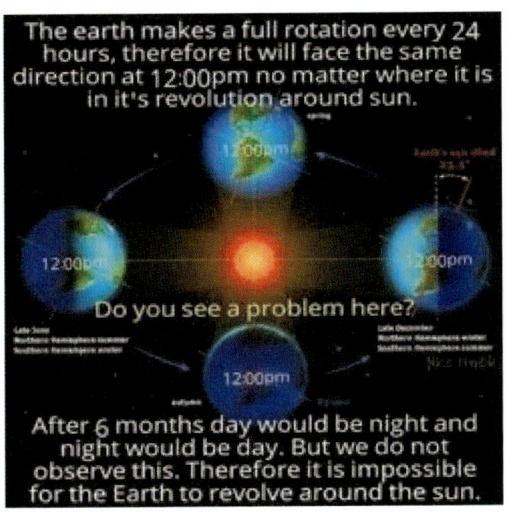

The Earth is said to revolve consistently every 24 hours around a Sun which is said to be stationary relative to Earth. After six months the spinning ball would have traveled almost 266 million miles and be on the other side of the Sun.

In six months, according to the heliocentric model, day and night would flip every six months, given a constant Sun and a rotating Earth, yet this never occurs proving, once again, Earth is not a spinning ball and/or does not rotate around the Sun.

Nonsensical Eclipse of the Mind

Another supposed proof of Earth's shape are the solar and lunar eclipses that occur periodically overhead. A lunar eclipse is said to occur when the Earth passes between the moon and the Sun, and the Earth's shadow obscures the moon or a portion of it. A solar eclipse occurs when the moon passes between the Earth and the Sun, blocking all or a portion of the Sun. An eclipse can be total, partial, or annular.

All three must be in near perfect alignment for one sphere to block the light from another sphere as you can see in the three lined up billiard balls.

When we view an eclipse, it is always above us. How can Earth be in alignment for an eclipse to occur with the Sun and the Moon *up* there, and Earth *down* here?

For the Sun's light to be casting Earth's shadow onto the moon, the three bodies must be aligned in a straight 180-degree syzygy, but as early as the time of Pliny, there are records of lunar eclipses happening while both the Sun and moon are visible in the sky. Therefore, the eclipsor of the moon cannot be the Earth/Earth's shadow and some other explanation must be sought since Earth is 'below' the Sun and moon above.

For modern heliocentric theoretical astronomy to validate eclipses they had to make the Sun 400x larger than the moon and exactly 400X farther away from Earth so it all would fit perfectly into their heliocentric model.

The moon is its own luminosity and is said by ancient cosmology to have its companion moon, Black Rahu. The dark moon, Black Rahu, passes in front of its binary Moon causing the moon phases. It should also be noted that ancient Chinese astronomers were able to predict eclipses as far back as 2500 BC without aid of modern day telescopes, instruments or the Nazi's and Freemasons who created NASA.

The UN Knows Flat Earth

The United Nations land was donated by the Rockefeller family in the 1940's on what was previously a slaughter house. The UN charter was ratified in on October 24, 1945, by the original 5 permanent members of the Security Council. Why would they choose a Flat Earth emblem for their flag?

Flags are symbols. Symbols rule the world, since one picture, or image, is worth 10,000 words, it is said. The UN is filled with Satanic symbology throughout. So why would so many International agencies purposely use a Flat Earth map in their logos? Or are they letting us with "eyes to see and ears to hear" the real truth through their symbology?

Bonus thought question: Since most of us use less than 1/5 of our brain capacity, the other 4/5 of our brain is subconscious and available for mind programming that is unfiltered by conscious awareness. Governments and business and science spends tens to hundreds of billions of dollars on mind brain mapping research. What do you think they have learned about how our minds work over the decades with all this time, energy and research?

Stars Seen Through the Translucent Moon?

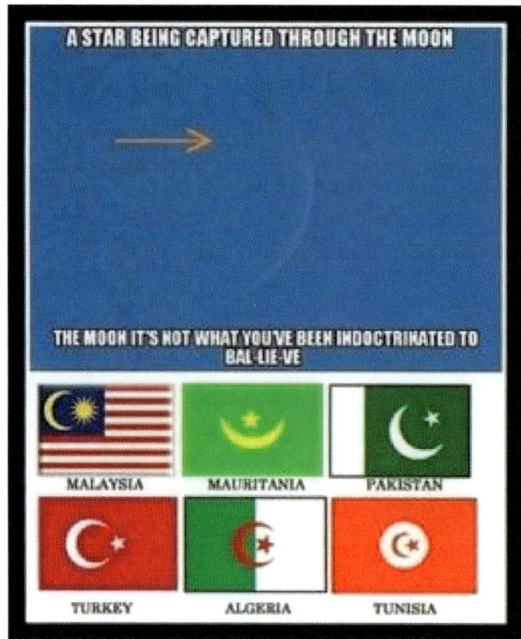

Crescent Moons appearing together with a star are a common feature of ancient culture's flags, including Sumerian, Persian and Egyptian iconography. The Sumerians used crescent shape of the Moon to be associated with the moon god Sin (Nanna) and the star with Ishtar (Inanna, i.e. Venus), often placed alongside the sun disk of Shamash. Many ancient cultures worshiped Sophia, the Goddess of Wisdom, also known as Isis, as the creator of the fecundity of all Life.

Using their own powers of observation at night they were able to see stars through the translucent crescent Moon. The flags of many ancient cultures depicted the star of Venus, or Aphrodite, held inside the crescent of the Moon. If the Moon is said to be some 238,000 miles from Earth, and the planets are millions of miles away, how can we see Venus through a crescent moon while just using our own observations through a telescope or high zoom digital camera?

The Coriolis Effect
The Spin on Logic and Reason

According to geophysics, the Coriolis effect is said to be caused when mass is moving in a rotating system that experiences a force (the Coriolis force) acting perpendicular to the direction of motion and to the axis of rotation. On the Earth, the effect is said to deflect moving objects to the right in the northern hemisphere and to the left in the southern hemisphere. The Coriolis effect is important in the formation of cyclonic weather systems, according to heliocentric theory.

The Coriolis Effect is said to cause sinks and toilet bowls in the Northern Hemisphere to drain spin opposite in the Southern Hemisphere, thus declared to provide proof of the spinning ball-Earth.

Sinks and toilets in the Northern and Southern hemispheres do not consistently spin in any one direction. Sinks and toilets in the very same household are often found to spin opposite directions, depending entirely upon the shape of the basin and the angle of the water's entry, not the supposed rotation of the Earth. And what is said to occur at the Equator, still water?

The Coriolis Effect cannot be replicated in a laboratory yet we are said that it occurs. When sailing south do we suddenly see a change in the way the water goes down the sink? No! Because, once again, the scientists tell us that size matters and only can be measured on the ocean of Earth. All effects of force one way or another occur by 1) the design of the bowl or sphere, 2) the power of liquid flow and 3) the direction from which the liquid enters the medium. Waterfalls would thus spiral perpendicular to the direction of motion, according to science, yet never do, thus disproving the Coriolis Effect theory.

Coriolis Effect: Earth spins at 1,000 mph West
To East yet oceans on one hemisphere will go one way
And go completely the opposite direction in the other
Hemisphere. Does this make any sense to anyone?

The Antarctic Anomaly
Closest to the Sun for Six Months of the Year
Yet Has 10,000 ft. of Snow and Ice?

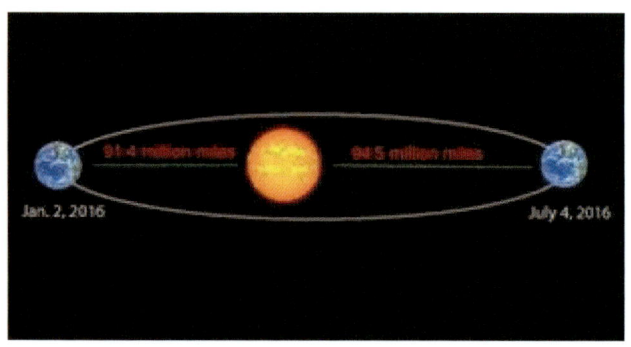 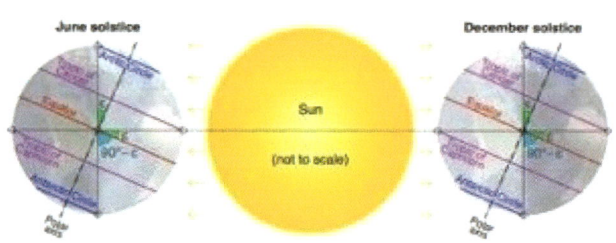

The Antarctic is the coldest place on Earth with snow 10,000 ft. deep and an ice belt 1200 miles thick. How is it that the Antarctic can stay so cold and frozen when it is closest to the Sun for six months a year and 3 million miles closer to the Sun during the elliptical perihelion?!?

How is it that the closest continents to the coldest place on Earth, Australia, South Africa and South America, are consistently the warmest places on Earth at the alleged "bottom" of the Earth?

If the Antarctic were closest to the Sun for six months of the year it should be acting more like the Arctic North pole which is becoming ice-free for the first time in over a million years? How can the Arctic be drastically melting while the Antarctic is solid ice and snow when both spend the same amount of time closest to the Sun during Earth's orbit?

Why was the Antarctic shut off limits by military declaration with the Antarctic Treaty System of some 13 countries in 1958, the exact same year NASA was founded? What did they wish to hide?

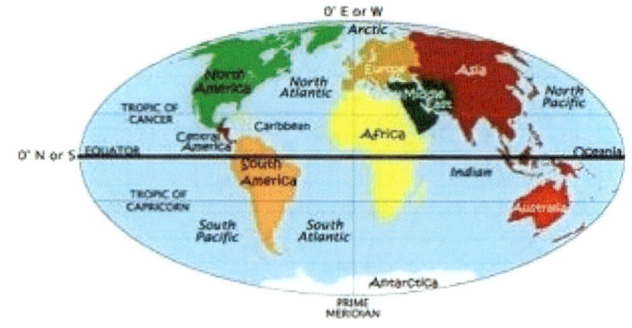 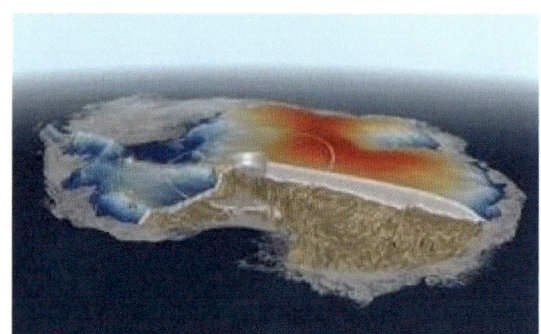

Ice/Snow thickness on Antarctica

Captain Cooks Circumnavigation of Antarctica
Only Works of a Flat Earth Map

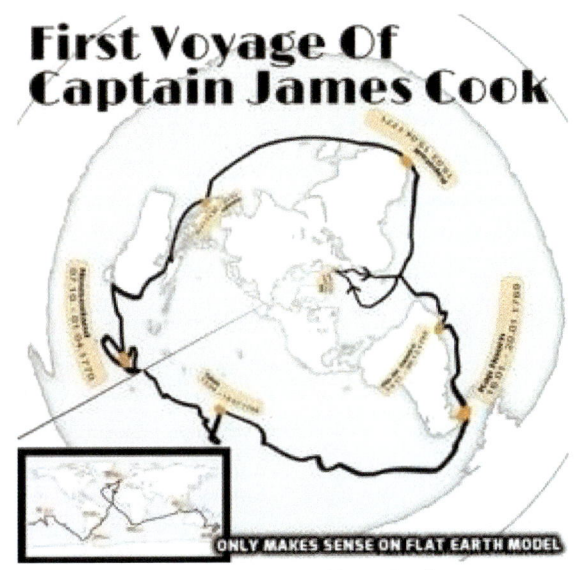

"In 1773, Captain Cook became the first modern explorer known to have breached the Antarctic Circle and reached the ice barrier. During three voyages, lasting three years and eight days, Captain Cook and crew sailed a total of 60,000 miles along the Antarctic coastline never once finding an inlet or path through or beyond the massive glacial wall! Captain Cook wrote: 'The ice extended east and west far beyond the reach of our sight, while the southern half of the horizon was illuminated by rays of light which were reflected from the ice to a considerable height. It was indeed my opinion that this ice extends quite to the pole, or perhaps joins some land to which it has been fixed since creation.'"

"On October 5th, 1839, another explorer, James Clark Ross began a series of Antarctic voyages lasting a total of 4 years and 5 months. Ross and his crew sailed two heavily armored warships thousands of miles, losing many men from hurricanes and icebergs, looking for an entry point beyond the southern glacial wall. Upon first confronting the massive barrier Captain Ross wrote of the wall, "extending from its eastern extreme point as far as the eye could discern to the eastward. It presented an extraordinary appearance, gradually increasing in height, as we got nearer to it, and proving at length to be a perpendicular cliff of ice, between one hundred and fifty feet and two hundred feet above the level of the sea, perfectly flat and level at the top, and without any fissures or promontories on its even seaward face. We might with equal chance of success try to sail through the cliffs of Dover, as to penetrate such a mass." ~ Eric Dubay, Flat Earth Conspiracy

"Yes, but we can circumnavigate the South easily enough,' is often said by those who don't know, The British Ship Challenger recently completed the circuit of the Southern region - indirectly, to be sure - but she was three years about it, and traversed nearly 69,000 miles - a stretch long enough to have taken her six times round on the globular hypothesis." —William Carpenter, *"100 Proofs the Earth is Not a Globe"*

"Upon the principle, as taught by Scripture and common observation, that the world is not a Planet, but consists of vast masses of land stretched out upon level seas, the North being the centre of the system, it is evident that the degrees of longitude will gradually increase in width the whole way from the North centre to the icy boundary of the great Southern Circumference. In consequence of the difference between the actual extent of longitudes and that allowed for them by the Nautical Authorities, which difference, at the latitude of the Cape of Good Hope, has been estimated to amount to a great number of miles, many Ship-masters have lost their reckoning, and many vessels have been wrecked. Ship-captains, who have been educated in the globular theory, know not how to account for their getting so much out of their course in Southern latitudes, and generally put it down to currents; but this reason is futile, for although currents may exist, they do not usually run in opposite directions, and vessels are frequently wrecked, whether sailing East or West." -David Wardlaw Scott, "Terra Firma"

Journey to Find the Dome
Operation High Jump and the Freemasons

Immediately upon returning on his first ever expedition to Antarctica in 1927, Admiral Richard E. Byrd (October 25, 1888– March 11, 1957) delivered some 27,000 documents and transcripts to the sponsor of his private sojourn, JD Rockefeller, the wealthiest man on Earth at the time. These papers have never been revealed to the public about what he discovered.

Admiral Byrd was a known 33-degree Freemason. In 1935 he sailed to Antarctica where he established the first Freemason Lodge, New Zealand Lodge 777. Immediately after WWII, in 1946, the US Navy high command unexpectedly, and in great hurry, launched the largest military operation ever down to Antarctica called Operation High Jump. The fleet departed Norfolk, Virginia on December 2, 1946. Admiral Byrd led the expedition of 30 ships and 4700 militarized soldiers (He also led some of the first expeditions to the North Pole.) The Antarctic mission had three task forces that were sent out in different directions. A British-Norwegian force and a Russian force, and some Australian and Canadian forces were also involved.

The rushed invasion of Antarctica was to last 6-8 months, yet ended, and the ships returned to home port, just 6 weeks into the expedition.

Two weeks before leaving on Operation High Jump, Admiral Richard Cruzen, commander of the flagship, Olympus, complained to Admiral Byrd he knew nothing about the purpose of their mission or supplies needed for his men just two weeks before departure. Even the commander of the fleet was not even aware of the mission, it was so secret. This was immediately following the end of WWII in 1945. Ships had to be recommissioned, including the rushed shakedown of the new aircraft carrier, the USS Philippine Seas, among the largest carriers of the time. The main force was divided into three groups. The Central Group comprised of the USS Mt. Olympus (communications); USS Yancey (supply); USS Merrick (Supply); USS Sennet (submarine); USCGC Burton Island (Icebreaker) and USCGC Northwind (icebreaker).

Operation High Jump went down with specially designed rockets, cranes and snow machines to traverse over the 150 + tall ice walls so they could venture deep into the Antarctic region. The mission failed completely as planes went off course and crashed when equipment failed likely due to the heavy electromagnetic fields closer to the edge for the dome. Many lives were lost, yet this was not reported to the public. The entire fleet had bugged out back to the states in six weeks as Admiral Byrd returned to tell fantastical tales of German Nazi anti-gravity UFOs down in the Antarctic that was reported in Chilean newspapers.

Admiral Byrd later spoke out in his autobiography, "*Alone*," about the 'hidden forces' that guided him throughout most of his missions. Just four months previous to Byrd's failed mission, the UFO Roswell incidents were being widely reported and the UFO meme was introduced into the public lexicon. Admiral Byrd mysteriously died in 1957. Was Byrd sent to find the Dome? If so, why was such a rushed expedition? Was equipment failure due to more intense fields of electromagnetic energy in the Antarctic?

1957 UN Forms Antarctica Treaty System;
Military Bases to Protect Uninhabited Ice?
...or to Keep the Dome from Discovery?

Donald Fagen, from the band, Steely Dan, wrote a song in the 1982 titled, I.G.Y. A song about the creation of the *"International Geophysical Year"* of study of the Antarctic regions from back in 1957. The Antarctic Treaty (ATS) was signed in Washington, DC on December 1, 1959 by the twelve countries whose scientists had been active in and around Antarctica during the International Geophysical Year (IGY) of 1957-58. The military enforced treaty went into effect in 1961. The ATS was set up at the same time that NASA was founded and is governed through Washington D.C. and the United Nations. As of 2015, there are 53 states party to the treaty, 29 of which, including all 12 original signatories to the treaty, have consultative voting status. Consultative members include the seven nations that claim portions of Antarctica as their national territory. The 46 non-claimant nations either do not recognize the claims of others, or have not stated their positions.

Travel to the Antarctic is highly restricted. Private air travel over it is strictly forbidden. Ships are only allowed to carry limited amounts of fuel said due to a spill that would damage the environment. Most tourist agencies are registered to similar spook agencies connected to Washington D.C. and there are still no live shots from space of the Antarctic, only Google images and maps.

Why would Donald Fagen write a song about the IGY in 1982 speaking about the fix being "in" and spandex jackets for everyone about a little known event occurring in 1957?

Get your ticket to that wheel in space
While there's time

The fix is in
You'll be a witness to that game of chance in the sky

You know we've got to win.
Here at home we'll play in the city
Powered by the sun
Perfect weather for a streamlined world
There'll be spandex jackets, one for everyone

Donald Fagen ~ I.G.Y.

  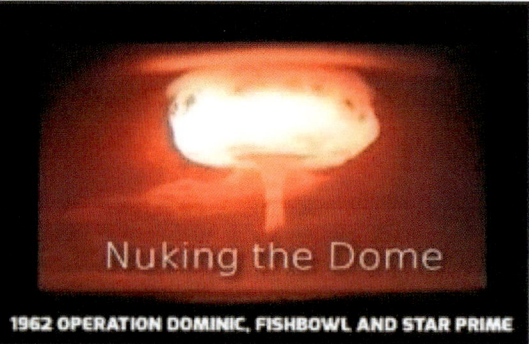

Nuking the Dome in 1962?

Operation Fishbowl was a series of high altitude nuclear tests in 1962 that were carried out by the United States as a part of the larger Operation Dominic nuclear test program. Operation Dominic was a series of 31 nuclear test explosions with a 38.1 Mt total yield conducted in 1962 by the United States in the Pacific Ocean.

At the exact same time as the US launches, the Soviet Union's K project nuclear test series was conducted with a group of 5 nuclear tests conducted in 1961-1962. The K project nuclear testing series were all high-altitude tests fired by missiles from the Kapustin Yar launch site in Russia across central Kazakhstan toward the Sary Shagan test range.

During the same year that NASA was founded in 1958, both Russia and the US signed a treaty banning thermos nuclear high-altitude testing. On April 12, 1961, Russian Yuri Gagarin became the first astronaut said to enter Earth's orbit in space. On May 25, 1961, in response to the Russian space launch, President John F. Kennedy announced to a joint session of Congress that the US was going to "catch up to and overtake" the Soviet Union in the "space race."

The US nuclear rocket explosions were conducted as part of Operation Dominic and included a series of high altitude tests known as Operation Fishbowl. These tests contained Thor missile launched warheads detonated at very high altitudes (30-248 miles) and said to evaluate the destructive mechanisms and effects of high yield explosions against ballistic enemy missiles. Several test failures occurred with missiles being destroyed in flight by range safety officers when electronics failed (Bluegill), when rocket motors malfunctioned (Starfish and Bluegill Prime), or when the missile veered out of control (Bluegill Double Prime). The Bluegill Prime test was particularly disastrous since the missile was blown up while still on the launch pad, requiring complete reconstruction of the demolished and plutonium contaminated Thor launch facility. The detonations of these missiles caused electrical blackouts hundreds of miles away, including Hawaii. No rocket got above 600 miles in altitude and had to be detonated after electronics went haywire. The James Van Allen radiation belt begins at 600 miles in outer space NASA tells us. Professor Van Allen accompanied Freemason Admiral Byrd on his last mission to Antarctica.

Could it be that these tests were an attempt to blow through the dome? Why did so many rockets go haywire and have to be detonated before accomplishing their tasks? Why did Russia and the US immediately reconstitute the High-Altitude Test Ban Treaty after the many failed rocket tests in 1962 by both sides? Is the Van Allen belts where our dome begins?

Founding Fathers of NASA
Freemasons and Nazi Scientists

C. Fred Kleinknecht, Chief Administrator for the command and service modules for all Apollo missions. After retiring from NASA, he was awarded the Sovereign Grand Commander of the Council of the 33rd Degree of the Ancient and Accepted Scottish Rite of Freemasonry of the Southern Jurisdiction. Was this his reward for pulling off the Apollo moon mission hoaxes? The 33rd degree Supreme Council, where Mr. Kleinknecht chaired from, is in the heart of Washington D.C. The Scottish Rite temple, said to entomb the first Supreme Council Commander, Albert Pike, is modeled after the descriptions of the Mausoleum of Halicarnassus, one of the seven wonders of the ancient world. The layout of the building also completes a pentagram star that connects the White House, Capitol Building and Washington Monument. George Washington, the Father of the United States was a Freemason. Many other Presidents were as well. (see Appendix II)

Thomas Keith Glennan (1958-1961) was NASA's First Chief Administrator. Dr. Glennan earned a degree in electrical engineering from the Sheffield Scientific School of Yale University in 1927. Following graduation, he became associated with the newly developed sound motion picture industry, and later became assistant general service superintendent for Electrical Research Products Company, a subsidiary of Western Electric Company, now General Electric. During his career, he was movie studio manager of Paramount Pictures, Inc., and Samuel Goldwyn Studios. Mr. Glennan joined the Columbia University Division of War Research in 1942, serving throughout World War II, first as Administrator and then, as Director of the US Navy's Underwater Sound Laboratories at New London, Connecticut. The perfect resume to help launch the fake space missions made in a Hollywood basement.

Wernher Magnus Maximilian, Freiherr Von Braun was a German Nazi, He designed and directed the Nazi rocket program during WWII. He later became the Chief Administrator for the Saturn rocket program at NASA for all the alleged Apollo missions to the Moon. He ran NASA's Saturn rocket program for 26 years straight. Von Braun was known as the "Father of Rocket Science." The US even created the "Von Braun Award" in his honor.

Von Braun was brought over by US military and high US government officials, along with over 3,000 other Nazi scientists, through Operation Paperclip that ran from the late 1940's up to the 1970's. Upon entry into the US, they were issued US passports and granted immunity from prosecution for their war crimes during WWII. Many scientists had practiced eugenics on POW's in Germany during WWII. While in the US, they helped run mind control experiments, like the secret MKULTRA experiments that effected thousands of innocent human lab rats.

Von Braun was a member of the Nazi party and the SS, (though they made him change to a suit for this picture). He was the rocket architect that designed the VI and VII "buzz" bombs that rained hellfire over England and other European allies and was said to be responsible for the killing of tens of thousands of innocent people towards the end of WWII. He was also suspected of perpetrating war crimes during World War II but much of his record in Germany during WWII has been white-washed.

NASA Fails

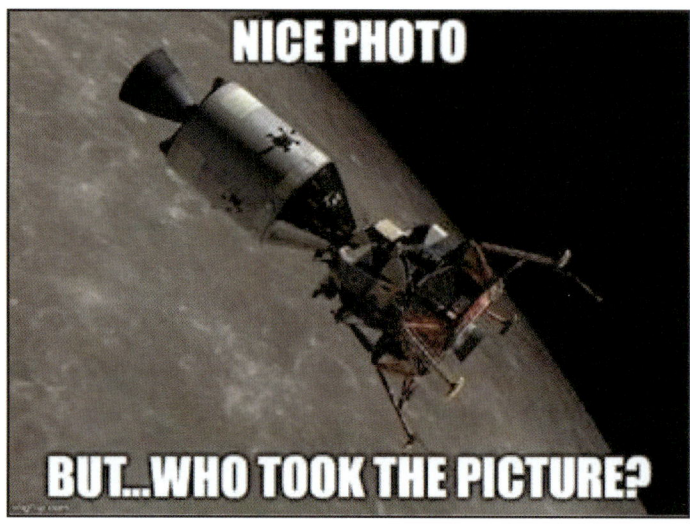

Since 1958, The National Aeronautical Space Administration (NASA), has been the single source of all news about space travel and modern day scientific astrophysics and cosmology.

NASA is, and always has been, a military operation disguised as public service to explore the realms of outer space. Apollo 11 was the first Moon mission, we are told. All TV feeds the world saw of the Moon missions came from one TV feed from NASA on time delay. All video footage of the Apollo missions are lost, according to NASA, all 11,000 reels, when NASA archivists erased them all so that they could be reused. This is the official story from NASA.

NASA admitted in 2006 that no one could find the original video recordings of the July 20, 1969, landing. In 2009 they announced that they were able to digitally restore some of the moon landing footage. This is the official story from NASA.

Movie director, Stanley Kubrick, is alleged to have assisted in the providing of movies for the faked Moon landings. Much of the film editing for the fake moon shots was done at the top of Lookout Mountain, Laurel Canyon, California at CIA photo-shop labs.

Starting in the early days of the Cold War (late 40's), the CIA began a secret project called Operation Mockingbird, with the intent of buying influence behind the scenes at major media outlets and putting reporters on the CIA payroll. Media assets will eventually include ABC, NBC, CBS, Time, Newsweek, Associated Press, United Press International (UPI), Reuters, Hearst Newspapers, Scripps-Howard, Copley News Service, etc. and 400 journalists, who have secretly carried out assignments according to documents on file at CIA headquarters, from intelligence-gathering to serving as go-betweens. The CIA had infiltrated the nation's businesses, media, and universities with tens of thousands of on-call operatives by the 1950's. CIA Director Dulles had staffed the CIA almost exclusively with Ivy League graduates, especially from Yale with figures like George Herbert Walker Bush from the "Skull and Crossbones" Society.

"Space may be the final frontier, but its made in a Hollywood basement" ~ Red Hot Chili Peppers

One Giant Blunder for Mankind

Or How NASA "Erased" All Moon Tapes From Their Archives

You can read the astonishing world news headlines from 2006. Just at the time digital imaging was taking apart the NASA photos, they were lost. The missing tapes were among over 700 boxes of magnetic data tapes recorded throughout the Apollo program which have not been found. On August 16, 2006, NASA announced its official search saying, "The original tapes may be at the Goddard Space Flight Center or at another location within the NASA archiving system" and "NASA engineers are hopeful that when the tapes are found they can use today's digital technology to provide a version of the moonwalk that is much better quality than what we have today."

Then, according to Wikipedia, they were found to be erased and copied over due to…wait for it…wait for it… Tape shortages in the 1980s, so they used the official only back up tapes of Apollo moon landings: "The researchers discovered that the tapes containing the raw unprocessed Apollo 11 SSTV signal were erased and reused by NASA in the early 1980s. It is claimed this was according to NASA's procedures because they were facing a major data tape shortage at that time."

Impossible Moon Photography

The Moon temperatures during the Apollo missions was over 200° F degrees in the Sun, we are told. This means cameras, even the best Nikon of the late 60's, would be tested like never before on Earth. Astronauts turned moonwalkers turned cameraman were recording the first ever pictures of the Moon and Earth from afar for time memorial of the epic journey. All in their self-contained heating, breathing and cooling life support systems with bulky moon boots and gloves to touch metal devices hundreds of degrees warm.

By NASA's own admission over 5,771 photos were taken, by 12 astronauts while on the moon during six moon missions. That is nearly 1,000 photos taken per their short one and two-hour moon walks outside their lunar model (LEM) during their Extra Vehicular Activity (EVA) on the moon's surface.

Here is the summation by photo-analyst, Jack White: Let's review how many pictures NASA says were taken:

Apollo 11……….. 121 Apollo 12……….. 504 Apollo 14……….. 374 Apollo 15……….1021 Apollo 16……….1765 Apollo 17……….1986

The time available to take photographs along with many other tasks like exploration, taking samples, setting up equipment, etc., was in their Lunar Surface Journal. In it they broke down the times for specific tasks.

Apollo 11……..1 EVA …..2 hours, 31 minutes……(151 minutes) Apollo 12……..2 EVAs…..7 hours, 50 minutes……(470 minutes) Apollo 14……..2 EVAs…..9 hours, 25 minutes……(565 minutes) Apollo 15……..3 EVAs…18 hours, 30 minutes….(1110 minutes) Apollo 16……..3 EVAs…20 hours, 14 minutes….(1214 minutes) Apollo 17……..3 EVAs…22 hours, 04 minutes….(1324 minutes)

Total minutes on the Moon amounted to 4834 minutes. That amounts to 1.19 photos taken every single minute on the Moon, while still doing all other activities on the Moonwalks. Let's look at those other activities to see how much time should be deducted from available photo time:

Apollo 11… Inspect LEM for damage, deploy flag, unpack and deploy radio and television equipment, operate the TV camera (360 degree pan), establish contact with Earth (including ceremonial talk with President Nixon), unpack and deploy numerous experiment packages, find/document/collect 47.7 pounds of lunar rock samples, walk to various locations, conclude experiments, return to LEM.

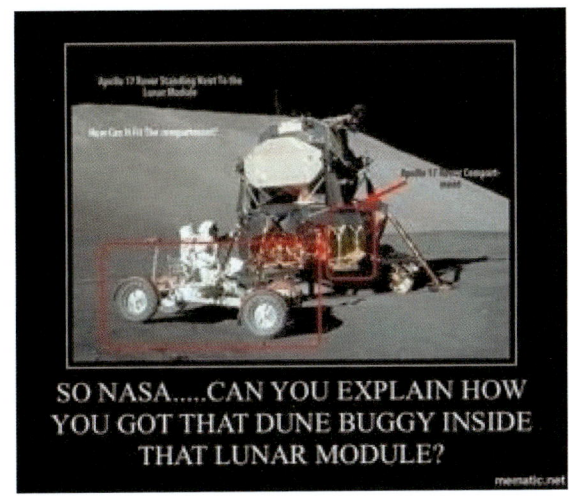

Apollo 12….Inspect LEM for damage, deploy flag, unpack and deploy radio and television equipment (spend time trying to fix faulty TV camera), establish contact with Earth, unpack and deploy numerous experiment packages, walk to various locations, inspect the unmanned Surveyor 3 which had landed on the Moon in April 1967 and retrieve Surveyor parts. Deploy ALSEP package. Find/document/collect 75.7 pounds of rocks, conclude experiments, return to LEM.

Apollo 14….Inspect LEM for damage, deploy flag, unpack and deploy radio and television equipment and establish contact with Earth, unpack and assemble hand cart to transport rocks, unpack and deploy numerous experiment packages, walk to various locations. Find/document/collect 94.4 pounds of rocks, conclude experiments, return to LEM.

Apollo 15….Inspect LEM for damage, deploy flag, unpack and deploy radio and television equipment and establish contact with Earth, unpack/assemble/equip and test the LRV electric-powered 4-wheel drive car and drive it 17 miles, unpack and deploy numerous experiment packages (double the scientific payload of first three missions). Find/document/collect 169 pounds of rocks, conclude experiments, return to LEM. (The LRV travels only 8 mph.)

Apollo 16….Inspect LEM for damage, deploy flag, unpack and deploy radio and television equipment and establish contact with Earth, unpack/assemble/equip and test the LRV electric-powered 4-wheel drive car and drive it 16 miles, unpack and deploy numerous experiment packages (double the scientific payload of first three missions, including new ultraviolet camera, operate the UV camera). Find/document/collect 208.3 pounds of rocks, conclude experiments, return to LEM. (The LRV travels only 8 mph.)

Apollo 17….Inspect LEM for damage, deploy flag, unpack and deploy radio and television equipment and establish contact with Earth, unpack/assemble/equip and test the LRV electric-powered 4-wheel drive car and drive it 30.5 miles, unpack and deploy numerous experiment packages. Find/document/collect 243.1 pounds of rocks, conclude experiments, return to LEM. (The LRV travels only 8 mph.)

When all other tasks are factored in as well on their missions, which most were said to be accomplished, here is the astonishing actual times that astronauts would have had to perform to get all the photos claimed.

Apollo 11……..one photo every 15 seconds Apollo 12……..one photo every 27 seconds Apollo 14……..one photo every 62 seconds Apollo 15……..one photo every 44 seconds Apollo 16……..one photo every 29 seconds Apollo 17……..one photo every 26 seconds

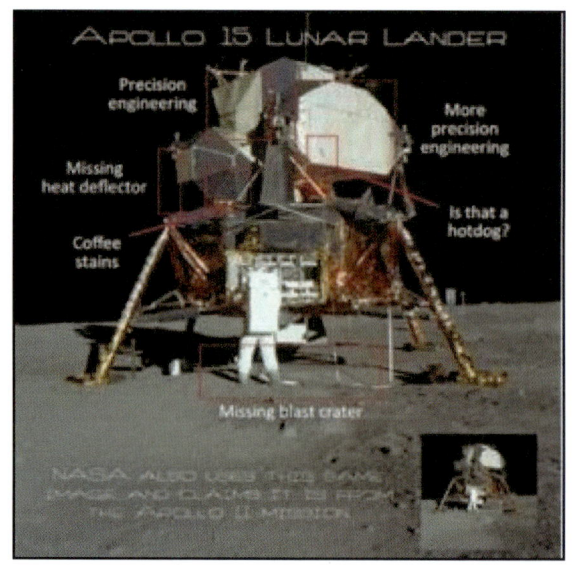

So you decide. Given all the facts, was it possible to take that many photos in so short a time? Any professional photographer will tell you it cannot be done. Virtually every photo was a different scene or in a different place, requiring travel. As much as 30 miles travel was required to reach some of the photo sites. Extra care had to be taken shooting some stereo pairs and panoramas. Each picture was taken without a viewfinder, using manual camera settings, with no automatic metering, while wearing a bulky spacesuit and stiff clumsy gloves.

The agency wants the world to believe that 5771 photographs were taken in 4834 minutes!

Hot Times on the Silvery-Gray Moon

		Daytime High			Nighttime Low		
		°C	°K	°F	°C	°K	°F
Mean Surface		107	380	225	-153	120	-243
Equator (0° Latitude)		122	395	252	-158	115	-252
Mid-Latitudes		77	350	171	-143	130	-225
Poles		-43	230	-45	-63	210	-81
Dark Polar Crater		-233	40	-387	-233	40	-387

"Exploration of the moon stopped because it was impossible to continue the hoax without being discovered. And of course, they ran out of pre-filmed episodes. No man has ever ascended much higher than 300 miles, if that high, above the Earth's surface. At or under that altitude the astronauts are beneath the radiation of the Van Allen Belt and the Van Allen Belt shields them from the extreme radiation which permeates space. No man has ever orbited, landed on, or walked upon the moon in any publicly known space program. If man has ever truly been to the moon it has been done in secret and with a far different technology.

The tremendous radiation encountered in the Van Allen Belt, solar radiation, cosmic radiation, Solar flares, temperature control, and many other problems connected with space travel prevent living organisms leaving our atmosphere with our known level of technology. Any intelligent high school student with a basic physics book can prove NASA faked the Apollo moon landings. If you doubt this please explain how the astronauts walked upon the moon's surface enclosed in a space suit in full sunlight absorbing a minimum of 265 degrees of heat surrounded by a vacuum... and that is not even taking into consideration any effects of cosmic radiation, Solar flares, micrometeorites, etc. NASA tells us the moon has no atmosphere and that the astronauts were surrounded by the vacuum of space."
—William Cooper, Former US Navy Intelligence

How is it possible for astronauts to go on moon walks for hours on end while their Lunar Escape Module (LEM) sat baking in over 200 F degrees' heat? How hot would that tin can be upon reentry? If the Sun is able to reflect so brightly off the Moon that it lights up the entire Earth from over a quarter-million miles away, how could the astronauts be able to see on the Moon? How could machines like moon buggies, and cameras all function so perfectly?

If Moon is in the vacuum of space, and there is little friction or drag to push off of, how did the LEM then create friction to lift off the Moon? And maneuver to dock with the orbiting command module? What aerospace engineer would ever design a flying machine in a pentagonal box shape?

They're Alive!?!?
Challenger Crew Members Found Alive and Gainfully Employed

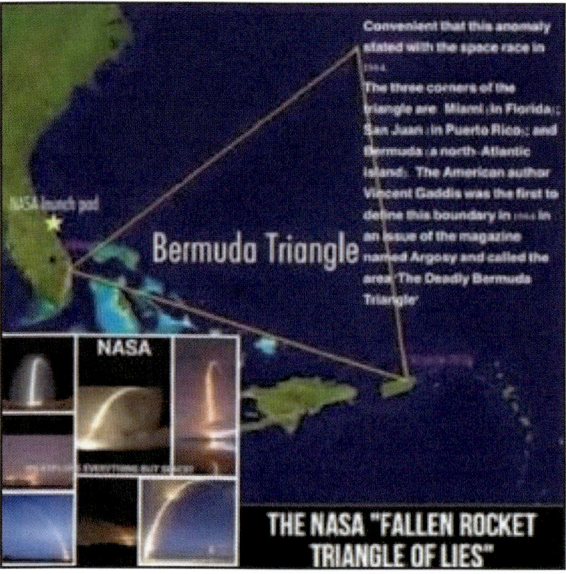

THEY LIVE! Lo' and behold some of those that allegedly died in the Challenger disaster have been found to be very much alive and gainfully employed, many under their own names. Two 'female NASA martyrs' appear to be still alive and in excellent health some thirty years since they exploded and fell from the sky. Ms. Judith Resnik and Sharon Christa McAuliffe seem to both be enjoying successful Law University positions and careers—while still using pretty much their own names—i.e., the names and surnames with which they briefly became international celebrities—as the NASA martyrs of the Challenger disaster.

Likewise, a fellow named Michael J. Smith —who bears a striking resemblance to NASA Challenger pilot Michael J. Smith (30-year time-lapse considered), is to be found alive and well, teaching at the University of Wisconsin, Madison. And fellow astro-not, Richard Scobee, who bears a striking resemblance (30-year time-lapse considered) to the Challenger's Commander, Richard 'Dick' Scobee, is to be found alive and well, as the CEO of a company called Cows in Trees. Likewise, both Claude Onizuka and fellow crew-member, Carl McNair have morphed into much look alike brothers named Ronald McNair and Ellison Onizuka, respectfully. What are the odds?

The Challenger crew did not blow up in space. It was all theater piped into everyone's heads by TV 'programming.' A non-reality show, faked as reality, while the billions of taxpayer dollars that payed for the show were transferred offshore to wealthy elite secret bank accounts, truth be told.

The Bermuda Triangle was created to explain strange phenomena that happens to ships and planes. The myth of the Bermuda Triangle was created to hide where the rockets, after launch in Cape Canaveral, Florida went to be ditched into the Sea.

ISS Space Station Does Not Exist
"But Hey, Wait a Minute, I Can See the ISS at Night!"

The ISS is said to be in orbit between 330 and 435 *km* (205 and 270 mi) above Earth. It is roughly the same size and dimensions as a 767-commercial jet liner. A commercial jet cruising altitude is around 30,000 feet or 6 miles above Earth.

How could we see the ISS more than 200 miles above us when we can hardly identify a jet flying at 30,000 feet, or five miles above in the sky if they are roughly the same shape and size?

Technology, especially in military applications, are over 25-100 years more advanced than is known to most. The military has long had planes that use ground based energy for perpetual flight without refueling. This was first proven to work as a ground based laser system by Tesla .

Project Blue Beam. Project Blue Beam was developed by DARPA (Defense Advanced Research Projects Agency) to project images into the sky using 3D laser projections of multiple holographic images. (Store bought laser pens can distract pilots thousands of feet up from the ground, so it is highly feasible that the military can project holograms on to, and off of the domed, or artificially domed sky at will.) Project Blue Beam has operational plans to deploy religious figures in space along with matching voices in our heads, as to simulate the visual  and audio second coming of Jesus Christ, Allah, Buddha, Krishna, or any other deity they wish. Voice inside the head, or V2K, as in Voice-to-Skull technology can be conducted wirelessly on any individual, anywhere in the world. (Google, "targeted individuals" to learn more about this most invasive technology.)

 The military has perfected advanced algorithmic Dwave super computer is feed data of physio-psychological particulars of individuals based on studies of the human anatomy, electric-bio functions, chemical and biological properties of the human brain. These computers are then loaded with all the languages of the human culture, along with their meanings (think Siri on steroids!). Using voice to head (also known as "voice of God") technology, they can speak any language into your head with a unique message from your personal God unique to you! At the same time they would be broadcasting a matching image into the sky of you most cherished religious icon. The Blue Beam Project may/could be used to fulfill age-old prophecies of many religions of a return of their Savior.

PBB would use the skies (60 miles up at the sodium layer of atmosphere) as a holographic movie screen. The purpose would be to unite all religions under one UNiversal Religion, as prophesized on the Georgia Guidestones located outside Atlanta, Georgia and commissioned in 1991, by R.C. Christian, covered further into this book.

Note that it was DARPA who gave us the internet in the mid-1990's.

Geostationary in Sync with Earth Spin;
So How Can We See the ISS at Night?

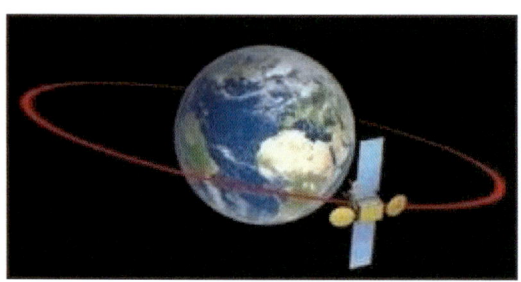

Geostationary Communications Satellites were first created in our minds by Freemason science-fiction writer Arthur C. Clarke. What supposedly was science fiction becoming science-fact just a decade later. Before this, radio, television, and navigation systems like LORAN and DECCA were already well-established and worked well using only ground-based technologies for all communications.

Wikipedia: "A geostationary orbit (GEO) is a circular geosynchronous orbit in the plane of the Earth's equator with a radius of approximately 42,164 km (26,199mi) (measured from the center of the Earth). A satellite in such an orbit is at an altitude of approximately 35,786 km (22,236mi) above mean sea level. It maintains the same position relative to the Earth's surface."

ISS Is Geo-synchronized with Earth. This means that the ISS space station cannot be seen unless you are directly underneath or near side to it due to the exact same synchronized speed it must keep to match Earths spin of 1,000 mph. A person standing on Earth either would always see the ISS if below it directly or nearly underneath it because it is matched to the ISS orbit above. Most on Earth could never even see the ISS space station because it is in near perfect synchronized orbit with Earth.

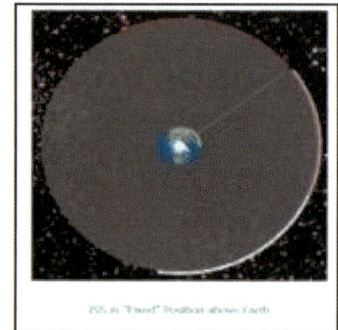

Some will taut the ISS.com website in that the solar panels, when struck at beneficial angles to the Sun, allow viewing of the ISS at night.

The spacecraft is traveling at 17,400 mph, temperatures in the thousands, tens of thousands of pieces of space debris in similar orbit and solar panels provide all the power to run the ISS space station. That's the story. The only thing that the powers who have controlled our minds for so long is our imagination. There is no space. It's a Google, or military duck decoy to fool the willing.

Communication Cable Routes Only Are Laid East and West

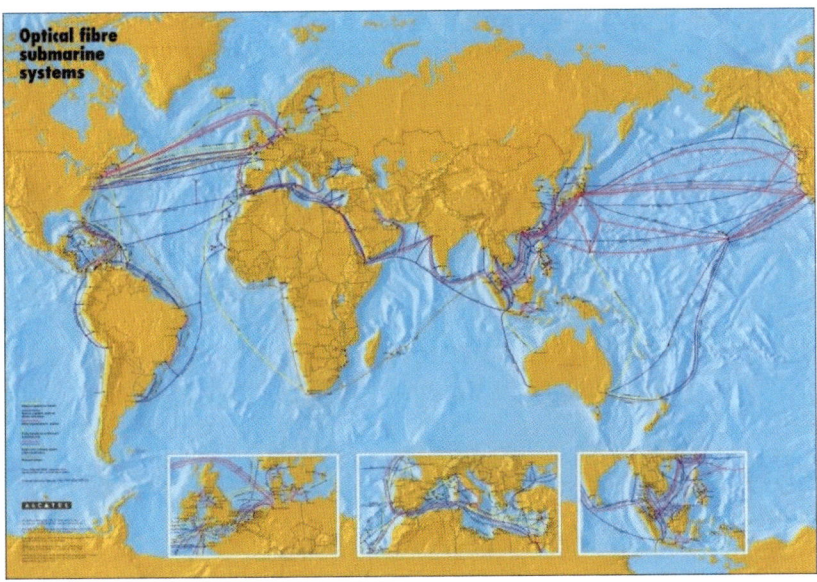

Cable routes are laid across the ocean floor East and West but not North and South. Fiber optic cable is expensive and corporations pay top dollar in the financial world for the fastest internet speeds possible. None of the hundreds of communication cables are laid through the much shorter routes over the Arctic and the Antarctic Circle. (For those that counter it is because of the cold and ice, please research the hundreds of miles of Alaskan pipeline!)

Fiber optic cable can cost hundreds of thousands of dollars for less than one mile of cable, and speed is essential. Why would they not then lay cable North and South if the route was so much shorter, they could save money and information would be delivered faster?

Fiber optic cables work perfectly on a Flat Earth model, hmmm!

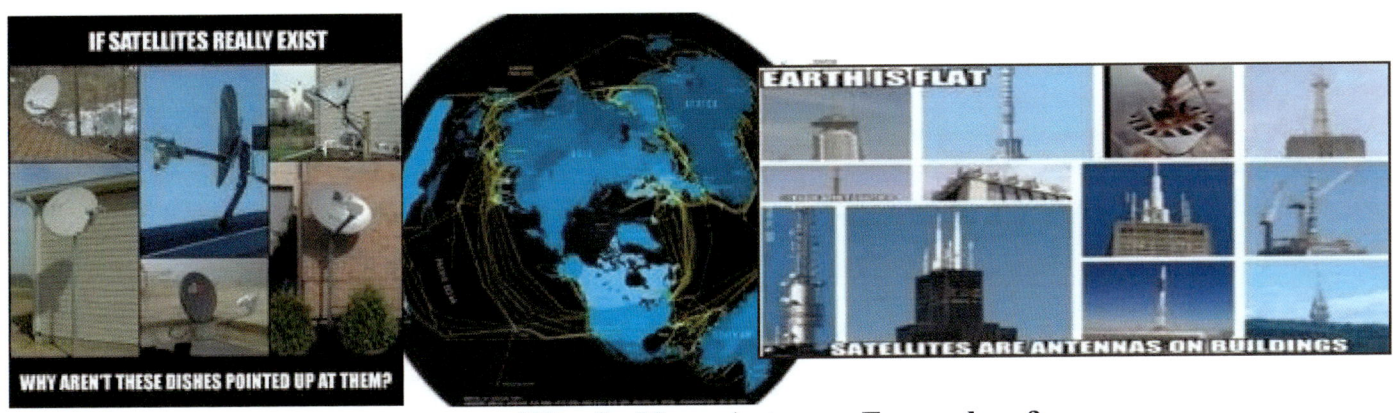

Why So Many Antennas Everywhere?

Why Are TV Dishes Pointed over the Horizon?

1. Satellites aren't used because they can't carry terabytes of data for less than a billion dollars per communication line.
2. The bandwidth available using a single fiber optic cable and a laser beam is much greater than you can get from a single satellite radio channel. This is due to the higher frequency and shorter wavelength of light compared to microwaves. The higher the frequency, the greater the bandwidth.
3. An undersea cable is a bundle of many fiber optic cables. Consider each fiber cable as a channel. You can have more channels, each with a higher capacity, than you can build radio channels into a satellite.
4. The up-links and down-links cost and putting the satellite in space is cost prohibitive.
5. The delay for satellite communications would be around 255 ms both uplink and downlink. For continuous traffic this not to a bad price to pay. But for burst traffic (like voice) you pay for the delay at each pause. The Rule of Thumb is 10 ms per 1000 miles so rule of thumb, to Europe on TAT–8 would be about 75 ms vs 510 ms for satellite.
6. Finally, you can fix a broken cable or cell tower. Once a satellite is broken it is disabled.

"Balloon-Powered Internet for Everyone"

"Project Loon balloons float in the stratosphere, twice as high as airplanes and the weather. In the stratosphere, there are many layers of wind, and each layer of wind varies in direction and speed. Loon balloons go where they're needed by rising or descending into a layer of wind blowing in the desired direction of travel. By partnering with Telecommunications companies to share cellular spectrum Google balloons enable people to connect to the balloon network directly from their phones and other LTE-enabled devices. The signal is then passed across the balloon network and back down to the global Internet on Earth." Google website

It is Google's goal to have internet provided everywhere and anywhere in the world, why would not they use satellites instead of balloons?

Global Positioning System Satellite Signal Simulator

"This invention relates to electromagnetic wave transmitters, and more particularly, to a transmitter for producing an output signal which simulates an orbiting GPS (Global Positioning System) satellite. The present invention will have many application and should, therefore, not be limited to those disclosed herein and in the drawings. However, the invention has been found to be especially useful when employed in connection with a test transmitter for GPS receivers.

In the past, test transmitters for GPS receivers have been employed to provide data and equivalent Doppler frequency for GPS receivers. Test transmitters for GPS receivers have been used as a method to reduce the cost associated with field tests. Unfortunately, such systems have been relatively complicated, and thus too expensive for practical commercial purposes. Accordingly, it is a principle object of the present invention to provide a relatively low cost GPS test transmitter while retaining high performance capability.

UMMARY OF INVENTION: Briefly, the present invention provides a new and improved apparatus for testing GPS receivers by simulating the signal and the equivalent Doppler frequency of an orbiting satellite. The test transmitter apparatus includes a computer which contains all the simulated mission maneuvers and orbit parameters of the satellites, a monitor for indicating the parameters of the apparatus, a keyboard for manual operation of the apparatus, and a signal generator that receives and decodes the information from the computer to emulate the appropriate satellites." US Patent # 5093800

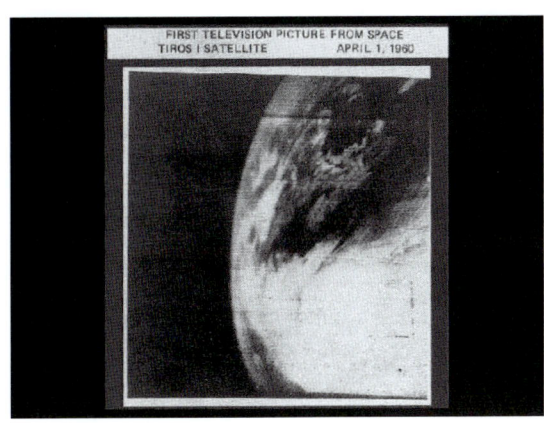

Sidebar: On April 1, 1960, NASA launched Tiros I

Microwave GPS Systems Are OTH Ground Based

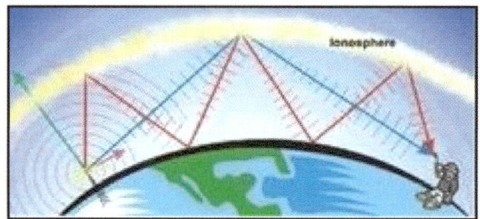 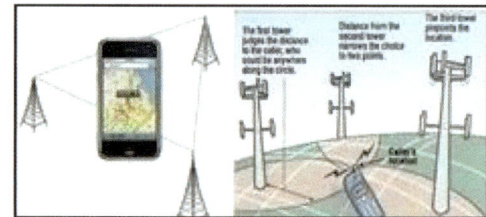

Since back in the 1940's when the Russians developed the Over-the-Horizon (OTH) based radar systems, wireless communication has been possible from ground based systems.

The most common type of OTH radar uses ionospheric reflection. Given certain conditions in the atmosphere, radio signals broadcast up towards the ionosphere will be reflected back towards the ground. After reflection off the atmosphere, a small amount of the signal will reflect off the ground back towards the sky, and a small proportion of that will reflect back towards the broadcaster.

The high frequency radio waves used by most radars are called microwaves and travel in straight lines. This generally limits the detection range of radar systems to objects on their horizon. If the target is above the surface, this range will be increased accordingly, so a target 10m (33ft) high can be detected by the same radar at 26km (16mi). Siting the antenna on a high mountain can increase the range somewhat, but in general it is impractical to build radar systems with line-of-sight ranges beyond a few hundred kilometers.

OTH radars use various techniques to see beyond that limit. Two techniques are most commonly used are shortwave systems that reflect their signals off the ionosphere for very long-range detection, and surface wave systems which use low-frequency radio waves. These systems achieve detection ranges of the order of a hundred kilometers from small, conventional radar installations.

For greater distances repeater stations are located to boost the signal along the horizon. Simple observation will show you how there are cell towers everywhere now. On buildings, fake trees, antennas, towers, etc. are all positioned to achieve greater distance from a higher position, just like the Loon balloons used by Google. *There are no satellites in space. There is no satellite triangulation of cell phone calls. Global Positioning Satellites or GPS do not exist.*

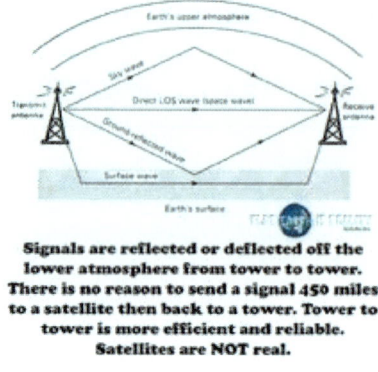

Signals are reflected or deflected off the lower atmosphere from tower to tower. There is no reason to send a signal 450 miles to a satellite then back to a tower. Tower to tower is more efficient and reliable. Satellites are NOT real.

NASA Is a Military Operation Inc.

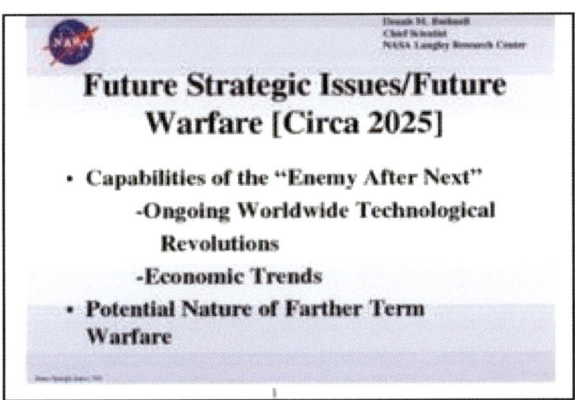 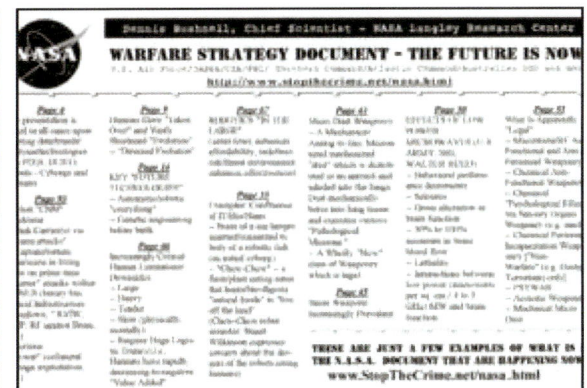

NASA is first and foremost a military operation disguised as a private corporation listed on Dun and Bradstreet's list of corporations. The above documents are taken from a slide show presentation that was said to be released inadvertently to the Internet in 2005 and passed around the alternative news media in 2013. NASA has not denied its authenticity. This 114-page slide presentation given by Dennis M. Bushnell, Chief Scientist at NASA's Langley Research Center, documents NASA's war strategies, including the use of highly advanced aerosol spraying technology used for mind control and human sterilization war plans.

Additional war strategies include spraying 'Smart Dust' containing self-replicating nanobots that are ingested then mechanically self-assembled inside our bodies and that can attach to our nervous systems. Geoengineer's (weather modifiers), use HAARP and Over-the-Horizon radar systems along with super computers that can literally read and overwrite commands to your brain and nervous systems. This highly secretive and advanced nanotechnology is currently being used on what is being called, 'targeted individuals'. Please Google "targeted individuals" to learn more about the victims of this wireless torture.

According to NASA's own warfare documents, 'humans have taken over' and 'vastly shortened evolution' (pg. 9) and there is a need to exploit the 'CNN Syndrome' to 'sink carriers via swarm attacks,' 'capturing and torturing of Americans in living color on prime time,' 'infrastructure take-down using IO/IW, EMP, RF (wireless mind control technologies) against the Brain.' (pg. 93). What has this got to do with space exploration!

On page 50 we read about the possibilities of "Gross alterations of brain functions using low power radiation" and the confluence of artificial intelligence (AI) technology of "Robotics in the Large" including Chew-Chew—a flesh/plant eating robot that hunts/bio-digests 'natural foods' to 'live off the land' (pg. 35). Other secret documentation for war against the people include the "*Report from Iron Mountain*," "*Full Spectrum Dominance of Space by 2025*," and "*Silent Weapons for Quiet Wars*" included in the Bibliography/Resource section of this book.

We are All Being Sprayed Like Lab Rats
Geoengineering aka Chemtrails

NASA is intimately involved in aerosol spraying operations that have been conducted over all our heads now for over five decades. FIVE DECADES! Geoengineering is rarely reported by mainstream media, yet this does not mean it is not occurring in our skies regularly. Some know these aerosol spraying operations as "chemtrails" however, this term is not recognized by science or government.

Few are aware that heavy metal toxic chemicals and smart dust, carrying nanoparticulate mind control technologies, are being sprayed into our skies regularly, with harmful known and unknown effects to all life.

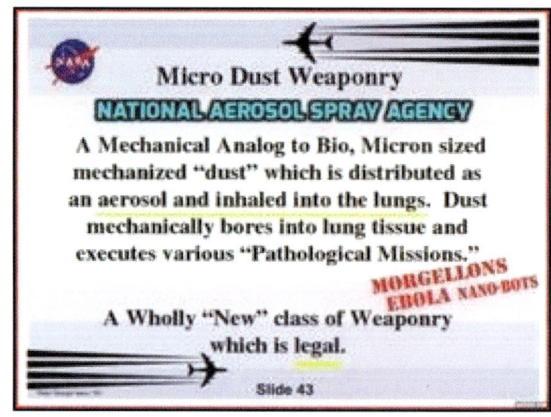

Since well before the Vietnam war of the 1960s and early 1970s, back to the days of William Reich and his Cosmic Orgone Energy (CORE) weather manipulator device, the military has been able to control the weather in many ways. They are so advanced with weather manipulation technology that there is virtually no more 'normal' weather. Today the weather manipulators can, and do, create Earthquakes, intensify hurricanes, lower the ionosphere for more acute communications for wireless mind control of targeted individuals as well as conduct aerosol spray operations dousing lithium for crowd control purposes or virus' to create illness known as "Chemtrail cough."

During the Arab Spring uprising in 2013, weather was geoengineered over Egyptian protesters to increase anger and rioting. Another aerosol inoculation of innocent people occurred during the Winter of 2015, in Southern Oregon. Well documented evidence proved Lithium, a mind drug, was sprayed over Southern Oregon during the protests to keep protesters calm and compliant. In 2016, many reports of overhead aerosol spraying came from the protestors in North Dakota Pipeline (NPL) protests.

As of 2016, Aerosol vaccinations are reported to being sprayed over Australia as well in the 5 boroughs of New York City, to 'prevent' the spread the Zika virus. Mr. Bill Gates 'has developed artificially intelligent (AI) mosquitoes that can carry virus' and is said to have tested this invasive technology in Brazil where the Zika virus is alleged to have originated from. And it is all perfectly legal. All without permission of the people, nor known was the health effects of such spraying on all.

In 1955 "Operation Drop Kick" purposely released infected mosquitoes on poor African American populations in Georgia and Florida as part of the much larger Tuskegee Operation that lasted from 1932–1972. In 1966, San Francisco residents were sprayed with Serratia marcescens bacteria to determine what biological weapons might simulate a germ-warfare attack. At the time, according to Rebecca Kreston for Discover magazine, it was 'one of the largest human experiments in history' and 'one of the largest offenses of the Nuremberg Code since its inception.' The Serratia bacteria has been known to cause meningitis, arthritis, wound infections, and transfers through dialysis, blood transfusions, cauterization, and lumbar punctures. Over the past 40 years there have been are over 50 US patents taken out specifically for geoengineering purposes that can only be used for aerosol spraying from aircraft.

I USED TO THINK THE TOP ENVIRONMENTAL PROBLEMS WERE BIODIVERSITY LOSS, ECOSYSTEM COLLAPSE AND CLIMATE CHANGE. I THOUGHT THAT WITH 30 YEARS OF GOOD SCIENCE WE COULD ADDRESS THOSE PROBLEMS. BUT I WAS WRONG. THE TOP ENVIRONMENTAL PROBLEMS ARE SELFISHNESS, GREED AND APATHY... AND TO DEAL WITH THOSE WE NEED A SPIRITUAL AND CULTURAL TRANSFORMATION - AND WE SCIENTISTS DON'T KNOW HOW TO DO THAT.

GUS SPETH

LIVE LEARN EVOLVE

A UN-Creation Story

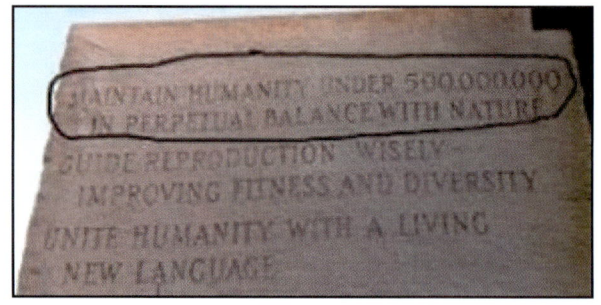

There is a very significant and important monument near Atlanta, Georgia in the United States unknown to most except conspiratorial theorists and deep state researchers. It was erected in 1991 by a person named R.C. Christian. He claimed to be representing a group of unnamed others who spent tens of millions of dollars to build what is known as the "Georgia Guide stones" or the "American Stonehenge."

Artisan craftsman from all around the world were summoned to carve 5 massive stones etched into granite with languages from ancient times to modern. Though millions of dollars were spent on the project, no one to this day claims ownership. These stone carvings were taken from the finest granite in the United States and located directly over a major geodesic energy lay-line of power. Like Stonehenge in England, these four monoliths with a crowing capstone were constructed and set to stand for centuries. However, unlike the smooth stones of Stonehenge, the guide stones contain many languages and words in what is apparently the groups version of the Moses ten commandments. They appear to be calling out their future agenda and plans for the entire world.

The stones are etched and written on all eight sides with the same ten commandments in eight different modern day languages, English, Hebrew, Russian, Swahili, Arabic Chinese, Hindi and Spanish. The much thicker heavy capstone that sits atop the four granite pillars are written in the ancient languages of Babylonian Cuneiform, Vedic Sanskrit, Classical Greek and Egyptian Hieroglyphics. Bottom line is this group of people have serious intentions as to how they wish to see the world become in the future.

Capstone in ancient languages "Let this be the Age of Reason."
The 10 commandments: The ten guides for a new Age of Reason are as follows:

1. ***Maintain humanity under 500,000,000 in perpetual balance with nature.***
2. Guide reproduction wisely — improving fitness and diversity.
3. Unite humanity with a living new language.
4. Rule passion — faith — tradition — and all things with tempered reason.
5. Protect people and nations with fair laws and just courts.
6. Let all nations rule internally resolving external disputes in a world court.
7. Avoid petty laws and useless officials.
8. Balance personal rights with social duties.
9. Prize truth–beauty–love–seeking harmony with the infinite.
10. Be not a cancer on the Earth—Leave room for nature—Leave room for Nature.

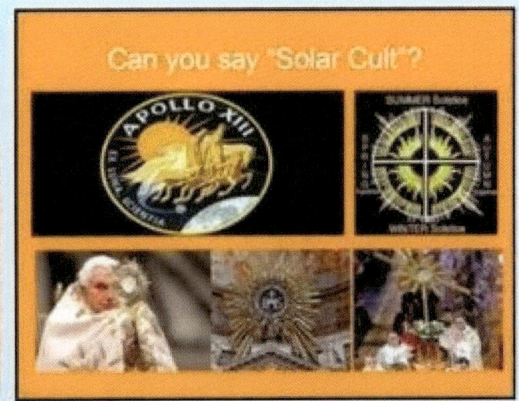

L.U.C.I.F.E.R "Large Binocular Telescope Near-infrared Utility with Camera and Integral Field Unit for Extragalactic Research." Mt. Graham Observatory, Southeastern Arizona. Built and Co-owned with the University of Arizona and the Vatican.

The Vatican; Owners of the Most Wealth
And the Most Telescopic Observatories
Over the Longest Time in Modern History

In all modern history, interference with science in the supposed interest of religion, no matter how conscientious such interference may have been, has resulted in the direst evils both to religion and to science—and invariably. And, on the other hand, all untrammeled scientific investigation, no matter how dangerous to religion some of its stages may have seemed, for the time, to be, has invariably resulted in the highest good of religion and of science.
—Andrew Dickson White, 1834

In the 1800s the relationship between science and religion became an actual formal topic of discourse, while before this no one had pitted science against religion or vice versa, though occasional interactions were expressed in the past. More specifically, it was around the mid-1800s that discussion of "science and religion" first emerged because before this time, "science" still included moral and metaphysical dimensions, was not inherently linked to the scientific method, and the term "scientist" did not emerge until 1834.

For the next 50-years the Royal Institute of Astronomy, in partnership with the Society of Jesus, also known as the Jesuits, began to sell the heliocentric theory as the mass public education system was being rolled out in Europe and then across the pond to the America's. The Jesuit scientists were the first in Latin America to spread the gospel of astronomy as well as to Far East. By 1750, 30 of the world's 130 astronomical observatories were run by Jesuit astronomers. Jesuits were often highly trained in languages and the sciences and were expected to use their technical knowledge, especially in the realm of astronomy, to win over rulers and elites around the world and convince them, by extension, of the superiority of the Catholic message.

Through owning all of the information coming from the heavens, along with NASA, the Roman Catholic Vatican has been able to steer and control the narratives of space travel and cosmology. Since the 16th century, Vatican Jesuits have traveled as far as China (1540), to promote astronomy and later, heliocentric theory to the world. These narratives have included controlled opposition stories of scientific breakthroughs by Copernicus, Newton, Kepler and Galileo who all had their findings later vetted, packaged and sold through the Royal Institute of Astronomy in England and then to NASA.

The Vatican also owns most of the early books of antiquity on astronomy as well as owning the most patents of anyone on celestial observatories. Strange for an organization who was adamant of a geocentric universe and that God created the world we live upon.

The hierarchal priests allegedly crucified and burned at the stake heretics like Bruno, and exiled Galileo, for their trying to prove that the Sun was center of our solar system, not the Earth. And the story goes that Copernicus waited until on his death-bed before publishing the work that changed everything we have been told about our relation to the universe, the book, *On the Revolutions of the Celestial Spheres*, which he dedicated to Pope Paul III in the preface.

Why Do You Think the Vatican Holds Their Most Sacred Day on Sun-Day?
"Do What Thou Whilst"

"So many plots have already been made against my life, that it is a real miracle that they have failed, when we consider that the great majority of them were in the hands of skillful Roman Catholic murderers, evidently trained by the Jesuits."
—President Abraham Lincoln speaking with Ex-Priest Charles Chiniquy.

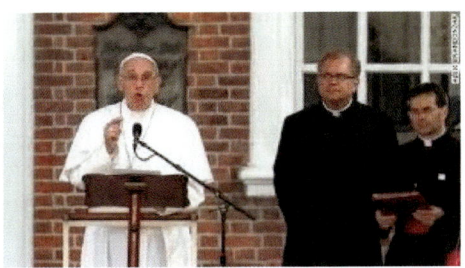

(*Pope Francis using Abraham Lincoln's lectern during visit to Philadelphia, PA*)

Directly after the trial and conviction of the Roman Catholics who murdered President Lincoln, all diplomatic relations by the United States was terminated with the Vatican until Ronald Reagan restored full diplomatic rights back to the Vatican. It was the US Congress that banned all diplomatic relations with the Jesuits after the trial and conviction of 9 Roman Catholics.

Some were hanged but John Surrat was given asylum in the Vatican after his conviction so the US Congress banned the Vatican from having an embassy in the United States. It wasn't until 1986 when President Ronald Reagan normalized relations with the Vatican.

Catholic John Surratt, one of the principal conspirators in Mr. Lincoln's assassination fled to the Vatican to avoid trial in the United States. Archbishop of New York. Other trial convicted co-conspirators were: John Wilkes Booth, Lewis Powell, David Herold, Michael O'Lauglin, John Surratt, Edman Spangler, Samuel Arnold, George Atzerodt. and Mary Surratt, all Roman Catholics.

~ excerpt from interview with Eric Phelps, Author of "Vatican Assassins"

"Cardinal Francis Spellman, ordered the assassination of John F. Kennedy directly after Kennedy demanded withdrawal from "Spelly's War" in Vietnam. Among many reasons, JFK was assassinated because he was going to withdraw troops from Vietnam and begun to print US currency not based on debt. It is of no coincidence that the first Pope, Francis, is the first Jesuit Pope in 600 years.

We know, on its face, that the Vietnam War was called "Spelly's War" - Cardinal Spellman's war. He went over to the warfront many times and he called the American soldiers the "soldiers of Christ." The man who was the Commander of the American forces was a Roman Catholic, CFR member, possibly a Knight of Columbus, I don't know, but he was General William Westmoreland.

So, Westmoreland was Cardinal Spellman's agent to make sure that war was prosecuted properly. And another overseer of Westmoreland was Cardinal Spellman's boy, Lyndon Baines Johnson. Lyndon Baines Johnson was a 33rd-degree Freemason. He was also part of the assassination, with J. Edgar Hoover, another 33rd-degree Freemason.

And Johnson went to Cardinal Spellman's death at St. Patrick's Cathedral, and the picture can be seen in Cooney's work The American Pope. So, Johnson was completely at the beck and call of Cardinal Spellman through Cartha DeLoach, the 3rd-in-control of the FBI. According to Curt Gentry, in his Hoover: The Man And The Secrets, DeLoach had a phone at his bedside direct to Johnson, and Johnson could call him anytime. DeLoach was a Knight of Malta, subject to Spellman.

Spellman wanted the Vietnam War, why? Spellman was controlled by the Jesuits of Fordham. Why did the Jesuit General want the Vietnam War? The people of Vietnam, the Buddhists, were unconvertible. They would not convert to Catholicism. They didn't need Rome. There had been a Jesuit presence in Vietnam for centuries, so it had been decided that about a million or so Buddhists would have to be "purged." They would later continue this purge of Cambodia, with Pol Pot, and the purge is yet for Thailand. It was a purging of Laos, Cambodia, and Vietnam of all these Buddhists, just like they purged the Buddhists of China with Mao Zedong, because Mao Zedong was completely controlled by the Jesuits. So, they wanted the Vietnam War.

The other thing is that Rome is in control of the drug trade. The Vatican controls all of the drug trade - all of the heroin, all of the opium, all of the cocaine, everything going around in Colombia. Colombia has a concordat with the Pope. A concordat is a treaty with the Pope. Hitler had a concordat. Mussolini had a concordat. Franco had a concordat. They want to set up a concordat here, which was the reason for Reagan formally recognizing the sovereign state of Vatican City in 1984. The greatest traitor we ever had was Ronald Reagan. So, they had a concordat. Colombia has a concordat. Do you think that drugs running out of Colombia, with a country that has a concordat with Rome, is not controlled by Rome? If Rome didn't want the drug trade out of Colombia, they'd end the concordat. The whole drug trade is run by high Mafia families out of the country of Colombia, subject to the Jesuit General.

And the Jesuit General ran the Opium trade, a couple of centuries ago, out of China. They ran the silk trade, the pearl trade. The movie Shogun is but a slight scratching of the surface of the Jesuit "black ships" that trafficked in all of this silk and pearls and gold and opals and everything they could pull out of the East, including opium.

The Vietnam War was to consolidate and control this huge massive drug-trade that would inundate every American city with drugs, being brought in by the CIA with their Air America, and then distributed by the Trafficante family throughout the United States - Santos Trafficante out of Miami.

We have the Mafia and the CIA working together in the drug trade. We have the Mafia and the CIA working together in the assassination of Kennedy. The first reason why the Jesuit General [at that time, Jean-Baptist Janssens] wanted Kennedy out of the way was because he was going to end the Vietnam War.

The second reason is, he wanted to end the reign of the CIA, because the CIA had betrayed him in the person of McGeorge Bundy, by not giving the cover to the Cuban patriots to retake Cuba from that Roman Catholic, Jesuit-trained, grease-ball bastard - he was a bastard, his father was a Nazi - Fidel Castro.

Kennedy was betrayed by the CIA at the Bay of Pigs invasion, which sacrificed all the patriots on the shores of the Bay of Pigs there, so Castro had no real opposition. This was the same tactic, used by the CIA and the KGB at the top, working together with Angleton controlling it, in the Hungarian Revolution, when the CIA fomented that revolution, and then betrayed all of those patriots into the hands of the Soviet army and KGB, which infuriated certain top CIA officials."

A controversial painting of John F. Kennedy's dying moments is on display at the Cathedral of St. Paul (the Vatican)

*****

All Presidents have their Roman Catholic handlers. Former President Obama's was Roman Catholic, Joseph Biden. One cursory look at President Obama's "Kitchen Cabinet" and you begin to understand who is really in power, command control….and has been for a very long time. Some of the key officials during President Obama's campaign and in his cabinet, include his top speechwriter, Jon Favreau, who trained at the College of the Holy Cross in Worcester, Massachusetts, the oldest Jesuit College in New England. Dan Pfeiffer, who was Obama's Deputy Communications Director during the campaign and continues in the same position at the White House, graduated from Georgetown University. His senior Military and Foreign Policy Advisor was Major General Jonathan Scott Gration, a fighter pilot whose master's degree is from Georgetown University, the oldest Jesuit institution in America. Most of the top brass of the Joint Chiefs of Staff graduated from the Edmund A. Walsh School of Foreign Service. President Bill Clinton graduated from Georgetown University and Donald Trump went to Fordham University, a Jesuit school. A Jesuit representing White House is once again guaranteed. The Governor of California, Jerry Brown, is a well-known Jesuit as is his likely successor, former Mayor of San Francisco, Gavin Newsom, who is a graduate of Jesuit school, Santa Clara University. Janet Napolitano, Head Regent for the University of California schools and former head of the Department of Homeland Security (DHS) graduated from Santa Clara University as well; just another coincidence theory?

MYSTIC NIGHTS OF THE ROUND FABLES is the coven of kindred Flat Earth mystics and philosopher friends all over the flat plane that resonate with this message. If you are a part of this movement, then you are part of the Order of the "Mystic Nights of the Round Fables".

Embedded within the white magic sigil is the initials MNOTRF in runic Nordic font (if you look close you will find them). The Sun and sun rays, the crescent shaped Moon as large half circles; The Antarctica Ice Wall forms Earth's outer circle. The Tropics of Cancer and Capricorn, Sun and Moon. 12 luminaries for the 12 zodiac signs as well as alchemists symbol for the divine masculine and feminine.

The Flat Plane and the firmament. The North star Polaris at the top. The flat and level ocean below. The luminaries circling about the pole star, Earth central Mt. Meru at the base. The Ascension Arrows connecting Earth with the heavens, pointing to the Awakening at large.
~ Watson Atkinson

Chapter 5

Reeducation and Reconnection

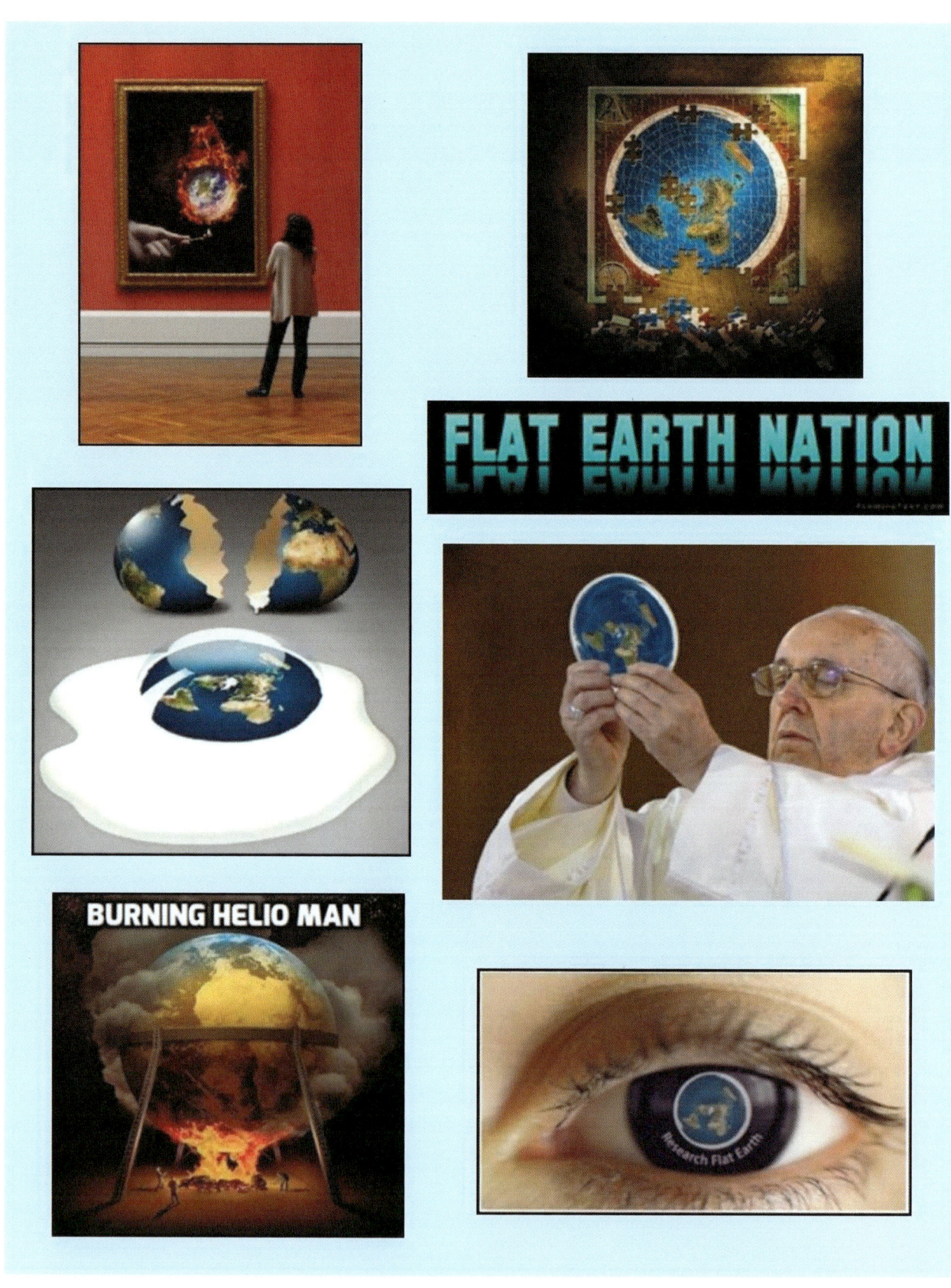

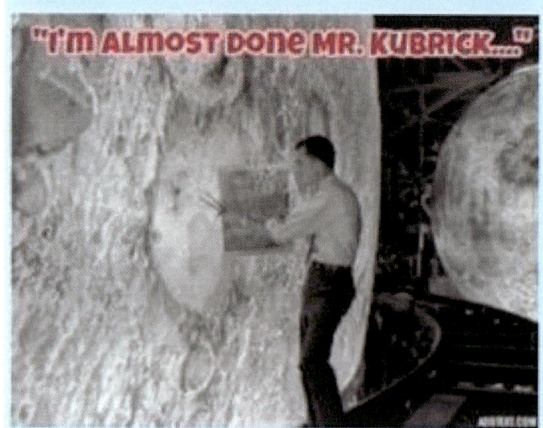
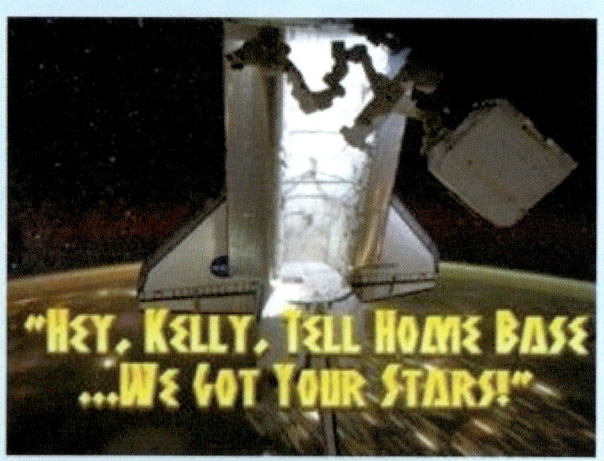
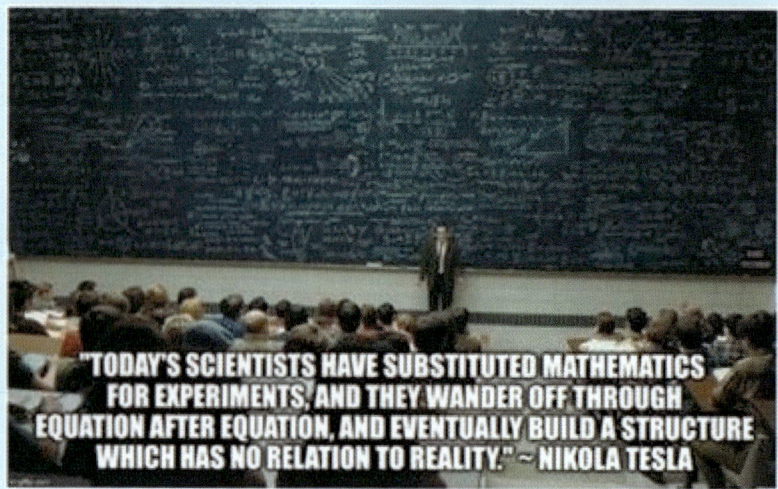

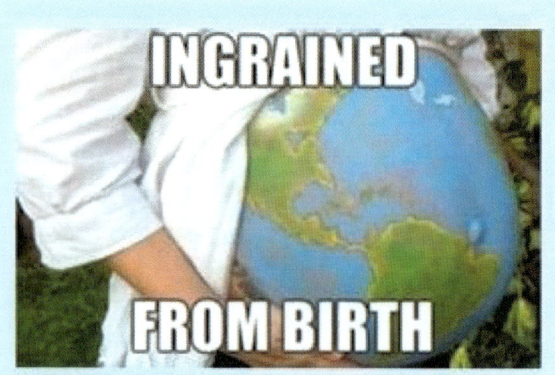

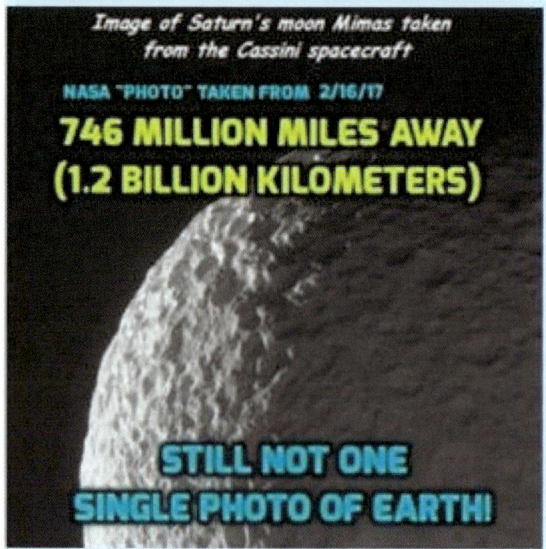
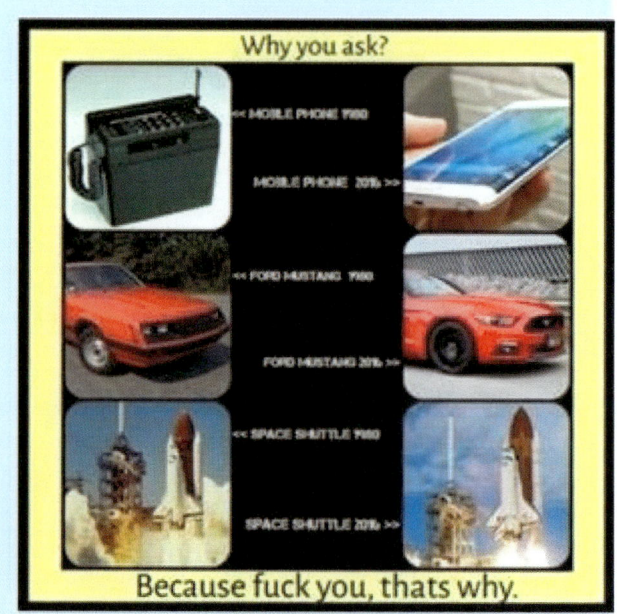

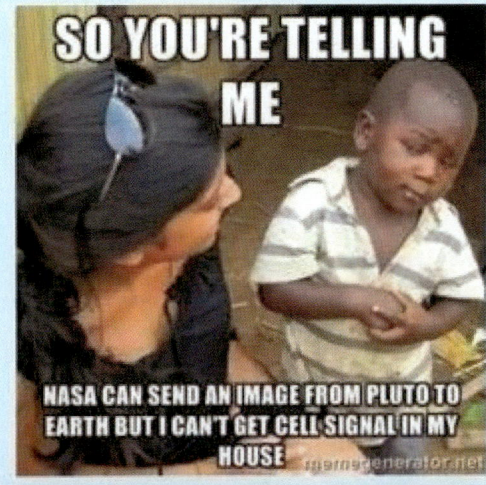

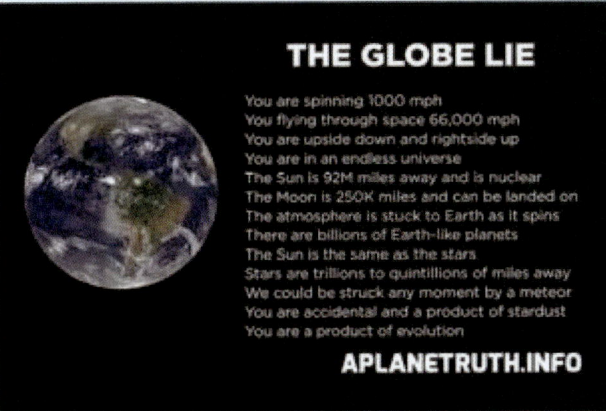

Epilogue

When I tell them that I'm doing fine watching shadows on the wall,
Don't you miss the big time, boy? You're no longer on the ball.
—John Lennon, Watching the Wheels

"We don't have time for a meeting of the flat Earth society." — Barack Obama

"They must have been founding members of the flat Earth society. They would not have believed that the world was round." — Barack Obama

"If I say that the world is round and somebody else says it's flat. That's worth reporting. But you might also want to report on a bunch of scientific evidence that seems to support the notion that the world is round." —Barack Obama

"Sixty years ago, when the Russians beat us into space. We didn't deny Sputnik was up there. We didn't argue about the science or shrink our research and development budget. We built a space program almost overnight and twelve years later we were walking on the moon." — Barack Obama

"I fly a lot, and I mean a lot. No one flies more than me. If the world was round, believe me I would know it!" — 45th US President, Donald Trump

"And we need, as responsible leaders, to take account of science – not some cockamamie ideological hypothetical, but science. And we need to make clear that those members of the flat earth society are on the wrong side of history." — US Secretary of State, John Kerry.

"In the Middle Ages people believed that the earth was flat, for which they had at least the evidence of their senses: we believe it to be round, not because as many as one percent of us could give the physical reasons for so quaint a belief, but because modern science has convinced us that nothing that is obvious is true, and that everything that is magical, improbable, extraordinary, gigantic, microscopic, heartless, or outrageous is scientific." — George Bernard Shaw, Author, Playwright (1856-1950)

It will be seen that my reasons for thinking that the earth is round are rather precarious ones. Yet this is an exceptionally elementary piece of information. On most other questions, I should have to fall back on the expert much earlier, and would be less able to test his pronouncements. And much the greater part of our knowledge is at this level. It does not rest on reasoning or on experiment, but on authority. And how can it be otherwise, when the range of knowledge is so vast that the expert himself is an ignoramous as soon as he strays away from his own specialty? Most people, if asked to prove that the earth is round, would not even bother to produce the rather weak arguments I have outlined above. They would start off by saying that 'everyone knows' the earth to be round, and if pressed further, would become angry. … This is a credulous age, and the burden of knowledge which we now have to carry is partly responsible. — George Orwell

"It seemed a Flat Disc with an upturned edge." — First man to reach the stratosphere in a balloon. Dr. Auguste Picard, 1931

"Einstein's theory of relativity is] a magnificent mathematical garb which fascinates, dazzles and makes people blind to the underlying errors. The theory is like a beggar clothed in purple whom ignorant people take for a king… its exponents are brilliant men, but they are meta-physicists rather than scientists."— Nikola Tesla, Inventor and Electrical Engineer (1856-1943)

"If the Government or NASA had said to you that the Earth is stationary, imagine that. And then imagine we are trying to convince people that 'no, no it's not stationary, it's moving forward at 32 times rifle bullet speed and spinning at 1,000 miles per hour.' We would be laughed at! We would have so many people telling us 'you are crazy, the Earth is not moving!' We would be ridiculed for having no scientific backing for this convoluted moving Earth theory. And not only that but then people would say, 'oh then how do you explain a fixed, calm atmosphere and the Sun's observable movement, how do you explain that?' Imagine saying to people, 'no, no, the atmosphere is moving also but is somehow magically velcroed to the moving-Earth. The reason is not simply because the Earth is stationary.' So what we are actually doing is what makes sense. We are saying that the moving-Earth theory is nonsense. The stationary-Earth theory makes sense and we are being ridiculed. You've got to picture it being the other way around to realize just how RIDICULOUS this situation is. This theory from the Government and NASA that the Earth is rotating and orbiting and leaning over and wobbling is absolute nonsense and yet people are clinging to it, tightly, like a teddy bear. They just can't bring themselves to face the possibility that the Earth is stationary though ALL the evidence shows it: we feel no movement, the atmosphere hasn't been blown away, we see the Sun move from East-to-West, everything can be explained by a motionless Earth without bringing in all these assumptions to cover up previous assumptions gone bad." — Allen Davies

"The ancient Chaldeans made very exact observations concerning the connection of human time-reckoning with the heavenly phenomena. They had a highly develop 'Calendar-Science'. Much that appears to us today as self-evident really dates back to the Chaldeans. Yet the Chaldeans were satisfied with a **mathematical picture of the Heavens which portrayed the Earth more or less as a flat disc**, with the hollow hemisphere of the heavenly vault arched above, the fixed stars fastened to it, and the planets moving over it." — Rudolph Steiner

"V for Vendetta"; A Movie made by, for and about the Roman Catholic Jesuits

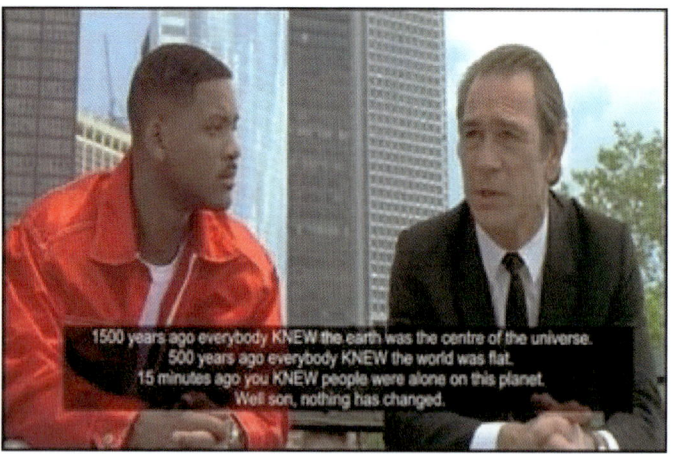

"The Real Men In Black"; A movie written by, for and about the Jesuits

"The Mission" A movie about Jesuits takeover of South America in the 1750's

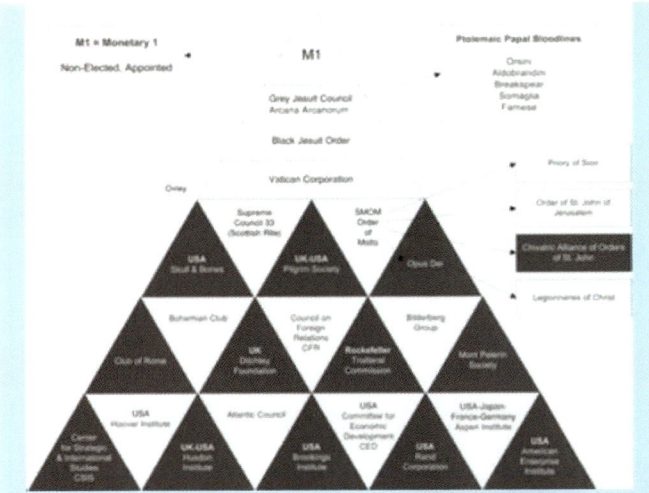

"My history of the **Jesuits** is not eloquently written, but it is supported by unquestionable authorities, [and] is very particular and very horrible. Their [the **Jesuit Order**'s] restoration [in 1814 by Pope Pius VII] is indeed a step toward darkness, cruelty, despotism, [and] death. ··· I do not like the appearance of the **Jesuits**. If ever there was a body of men who <u>merited eternal damnation on earth and in hell</u>, it is this **Society of [Ignatius de] Loyola**."

John Adams (1735-1826; 2nd President of the United States)

John Adams

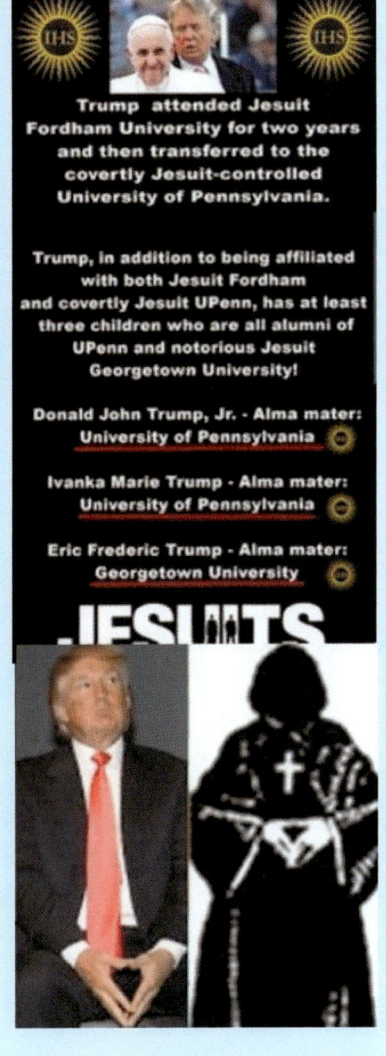

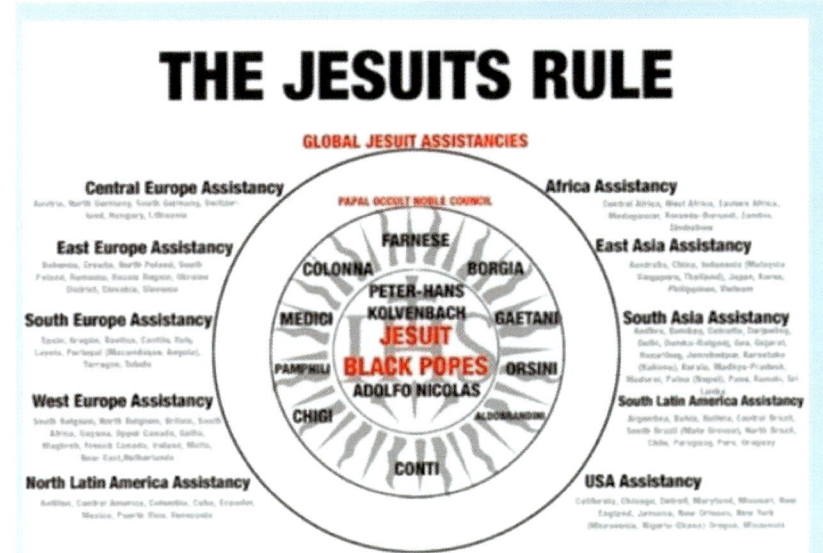

Appendix I

THE RULERS WHO MAKE THE RULES

Order Out of Chaos

In the United States, there is said to be over 16 million laws, codes, statutes, regulations, violations, infractions, ordinances, etc. that make up our legal system at the Federal, State and local levels, yet ignorance of these 16 million plus laws is no excuse, in the eyes of the courts. Most who live upon this Flat Earth world already live under a One World governance. This is simple to understand because no matter where we go in the world our personal debt follow us everywhere. There is only one central bank, the Bank of International Settlement (BIIS), domiciled in the protected country of Switzerland that was never invaded during WWI and WWII, and clears for all other sovereign country banks with only a few exceptions. As of 2001, there were only five countries left who are were not members of the BIIS; North Korea, Syria, Iran, Libya, Iraq and Afghanistan. Now, there are only three left, which the US is leading invasions in two as of 2016.

Additionally, the Federal Reserve bank of the US, which prints money, at interest for the US government, and makes the most money when more debt is incurred, is no more 'Federal,' than the corporation, Federal Express.

The word 'government' is derived from the Latin verb, '*Guvenere*' which means to rule or control. The Latin noun '*mentis* means 'mind.' The word 'government' literally means mind control.

The Roman Catholics, who before books being published in language other than Latin, controlled all narrative stories of our history. After Caesar pillaged and burned down the Great Library of Alexandria in 48 BCE, the Vatican was then free to write their own history, and thus what they wanted us to beLIEve, not what was true, or historically factual. The mass archives under the Vatican contain much of our world's true history but is inaccessible to all but those in the high powers of the Vatican. Heroes, like Father Alberto Rivera, who had access to the Vatican vaults, disclosed some of the truths hidden and it cost him his life.

The red cross, seen on the British and Swiss flags, is the same emblematic red cross that sailed on the ships commanded by Christopher Columbus, who was the real first American terrorist, if you were to ask the Native American Indians. The red cross is also the symbol for the Ku Klux Klan as well as the military arm of the Roman Catholic Jesuits, the Knights of Malta. The Rothschild's, some of the wealthiest Court Hofjuden Jews in the world today, changed their name from Bauer back in the 1700s. Roths means '*red*' and childs means '*shield*;' the red cross and the shield. The Rothschild's overtly are one of the richest families in the world today. It is also of no coincidence that Switzerland created the International Red Cross, so those in power could transport whatever they wanted across borders during World Wars. The red cross is one simple way to understand how symbols are more powerful, and rule our world, over laws and words. Symbols work directly into our unconsciousness and thus can be implanted into our minds without detection. The powers in charge have known this fact for centuries.

Ask yourself why, after the US War of Independence in 1812, the founders of the new country chose the exact same colors as the hated and defeated British, who they had just fought to the death where Ben Franklin challenged his fellow patriots to 'hang together or hang separately.' Does this make any sense to anyone? Also, the US flag, that Betsy Ross did not design, is nearly the exact same flag design as the infamous East Indian Tea Company, who were owned by the hated British.

According to American history, this was the event that set off the American Revolution. It is no coincidence that many other country flags are also red, white and blue, like Russia, Chile, French, Australian flags, and many others. Why when each country could choose any other colors in the spectrum?

Most of those who live in the United States have pledged, at one time or another, their allegiance to the US Constitution and American flag, not to a living being. Truth be told, not sold, the US Constitution was founded by the richest, white land-owning men of the time. When they finished with their 'We the People' document, the compact, or contract only included those who owned land, were rich, white and male. It would take almost another century before women, African American slaves, Native Americans, and non-land owning white folks to be legally allowed to vote.

Yet, most will cite the US Constitution, *just a piece of paper*, as proof of their 'unalienable rights' while living in an ever-increasing police state where we now have US President Donald Trump openly calling for stop and searches of any and all of those living in the US. For the record, Rights are not given by government, or written on a piece of paper as 'proof.' Rights are unalienable and inherent, not granted to us.

Few Americans even know the date or circumstances for founding of the Federal Government in Washington D.C. where the US President, Congress and the Supreme Law of the Land rule from (The Organic Act of 1871, February 23, 1871). This is when Congress created the USA Corporation and with the passage of the 14th amendment, did not free the slaves, as we were sold, but put us all under corporate rule of law on our citizen-ships. Our currency and court systems are still based on Roman Latin Law. One needs a Latin dictionary to decipher English law, yet few even know what the word 'Latin' even means or where the word is derived. Our money, or 'one-eye', system of exchange is all based on Roman Catholic laws created with the first Cestui Que Vie trusts by Pope Boniface, VII in 1306. This is when he and the Vatican Congress declared the Pope to be the trustee of all of Jesus Christs' holdings on Earth, including the reaping of all lands and souls.

These trusts are still in place in our western legal systems of today. BAR, as in BAR attorney, which most US attorneys belong to, stands for British Accredited Registry, a corporation domiciled in the City of London, England. Therefore, US lawyers who are BAR members, are registered to the Crown of England, whose corporate papers are registered away from any country law in Puerto Rico. Civil law is based on the Law of the Seas or Maritime Law, so to keep humans "ungrounded" and not under the "law of the land."

This is also why in court we go to into court on our citizenship, partnership, relationship, ownership, membership, leadership, friendship, etc. And why, in courtrooms in the US, you will see the American flag with yellow fringes and a tassel, and why we have *bailiffs* and enter the courts through a gate, called a *docket*.

Our monetary system is also derived from the Romans and Maritime Law of the Sea, that Romans have controlled since the days of Alexandria and Cleopatra. This is why we put our money in a *river* bank, with our *currency* (current-sea) holding our *deposit* slips while we account for our money (a piece of paper with ink on it) in *liquid* and *frozen* assets. It is not coincidence that the Vatican Roman Catholic Church is also known as the Holy See(a)! Through our willful ignorance, they have created a legal fictitious person in a contractual obligation through 'corps-orations,' meaning that we are all considered dead human 'resources,' or chattel (cattle) in the eyes of the law.

This is why government refers to us as 'human resources.' Your birth (berth) certificate was created, and only the mother's surname is used, if you were born in a hospital. The hospital is paid by the US government to register your baby at birth. 'Regis' means 'crown' and 'ster' means 'to enroll' in their corporation when you sign your name to a document (*dock*umet). The US birth certificate includes a registered securitized number that is pledged as collateral to the banksters since the time when the US declared bankruptcy in 1933. That ever-increasing debt has never been paid off, only accrued. After WWII, the bankers had to have collateral for the massive US debt incurred, so they securitized all of us at birth. The more debt we, and the governments incur, the more indebted servitude we all owe to the Court Hofjuden bankers and their handlers, the Vatican.

This is also why your name is in all capital letters on your driver's license, social security card, bank statements, etc. Every time you sign your name, you are signing a contract. This puts you under Vatican, City of London and Washington D.C., rulership. Maritime law is a system created and owned by the Roman Catholic Church whose Knights of Malta military has taken blood oath vows to: 'exterminate all heretics and Christians from the face of the Earth.' The Georgia Guidestones, outside Atlanta, Georgia are stone testaments to this fact to "keep humanity at 500 million people, down from our current 7.3 billion people said to inhabit Flat Earth today.

9-11-01; Right on the Money!

New $100 "Gold" Bill; "Abolish the Government"

Why did the US Treasury, in 2013, put the 2nd paragraph of the Declaration of Independence on the new $100 bills, including the words to "abolish government," along with a gold feather and ink well?

"That whenever any Form of Government becomes destructive of these ends, it is the Right of the People to alter or to abolish it, and to institute new Government, laying its foundation on such principles and organizing its powers in such form, as to them shall seem most likely to effect their Safety and Happiness."

Appendix II

SECRET SOCIETIES WHO RUN the WORLD

It is simply incredible how the so few, and very, very powerful, have kept secret their affiliations to so many secret societies from public knowledge and discourse for not only decades, but centuries. This is only a partial list of the true powers behind the power we see on social media today. And these secret societies answer in one way or another to the blacked robed Jesuits and their Knights of Malta military henchmen who report to the 13 Satanic family bloodlines of Europe.

* The Council of 13
* The Council of 33
* Secret Chiefs / Great White Lodge / Great White Brotherhood
* Order of the Quest
* Mothers of Darkness
* Moriah Conquering Wind
* Supreme World Council
* The Committee of 300 / The Olympians
* Old (Venetian) Black Nobility
* S∴S∴ / Third Order of the Silver Star / the Abyss
* The Bilderberg Group / Committee of 500
* A.A. / the Arcana Arcanorum
* The R∴C∴ / Order of the Rosy Cross
* Order of Palladium / New and Reformed Palladian Rite
* Skull & Bones Society / The Order / Brotherhood of Death / Chapter 322
* A∴A∴ / Arcanum Arcanorum / 'Argenteum Astrum' / Astron Argon
* A.P.R.M.M. / Ancient and Primitive Rite of Memphis-Misraïm
* The Round Table / Rhodes-Milner Round Table / The Group
* Palladian Order of Skull and Bones / Chapter 592
* The G∴D∴ / Hermetic Order of the Golden Dawn
* Fraternitas Saturni / Brotherhood of Saturn
* Order of the Trapezoid
* Grand Orient de France
* The Red Brotherhood
* Hell Fire Club(s)
* Frankfurt School
* The Pilgrims Society
* Le Cercle / The Circle
* Supreme Council / Mother Council of the World
* O.T.O. / Ordo Templi Orientis / Order of Oriental Templars
* S.M.O.M. / Order of St. John / Sovereign Military Order of Malta
* The Shrine / Ancient Arabic Order, Nobles of the Mystic Shrine
* S.R.I.A. / Societas Rosicruciana in Anglia
*Ordo Byzantinus Sancti Sepulchri (OBSS).
* Scottish Rite Freemasons

* York Rite Freemasons
* Order of the Garter
* The Temple of Set
* The Bohemian Club
* Order of The Hammer
* Order of Nine Angles
* The Black Brotherhood
* Scroll and Key Society
* Illuminates of Thanateros
* B'nai B'rith / B'nai Ha'Nephilim
* Ancient Mystical Order Rosae Crucis
* Ecclesia Gnostica Catholica / Gnostic Catholic Church
* Royal Institute of International Affairs
* Council on Foreign Relations
* Trilateral Commission
* U.S. Mafia Council
* The 1001 Club
* Club of Rome
* JASON Group
* Quill & Dagger
* RAND Corporation
* Lucis Trust / Lucifer Trust
* The MITRE Corporation
* British Royal Society
* Knights of Columbus
* Share International
* Societas Rosicruciana
* The Bridge to Freedom
* Phi Beta Kappa Society
* Tavistock Institute For Human Relations
* Muslim Brotherhood
* The Fraternity of The Rose Cross
* The Vrill
* Thule Society
* The Babylonian Brotherhood
* The Black Order
* The Brotherhood of Life and Death
* The Council of 10
* Order of The Green Dragon
* Ordo Lapsit Exillis
* Black Hand
* Shriners
* Opus Dei
* Order of the Garter

* Order of the Golden Dawn
* The Most Ancient and Most Noble
* Order of the Thistle Knight
* The Most Honourable Order of the Bath Knight/Dame Grand Cross
* The Most Distinguished Order of Saint Michael and Saint Order St-Michael St-George
* The Distinguished Service Order Companion
* The Royal Victorian Order
* The Order of Merit Member
* The Imperial Service Order
* The Most Excellent Order of the British Empire Knight/Dame Grand Cross
* The Order of the Companions of Honour
* Order of Hospitallers
* Order of the Church of the Holy Sepulchre
* Alfalfa Club
* Highlands Forum
* Prior of Scion
* Teutonic Knights
* Order of the Golden Fleece
* Fabian Society
* Theosophical Society

Here is a short partial compiled list of secret societies members. Listed are 33° Scottish Rite Freemasons, Freemasons, Knights of Malta and Skull and Bones members only.

US Presidents (since 1900)

Theodore Roosevelt – 33° Freemason
William Howard Taft – Freemason – Skull and Bones
Warren G. Harding – Freemason
Franklin D. Roosevelt – 33° Freemason
Harry S. Truman – 33° Freemason
Dwight D. Eisenhower – Knight of Malta
Lyndon B. Johnson -33° Freemason
Gerald Ford – 33° Freemason – member of JFK Warren Commission
Jimmy Carter – 33° Freemason
Ronald Reagan – 33° Freemason – Knight of Malta
George H. W. Bush – Freemason – Knight of Malta – Skull and Bones
Bill Clinton – 33° Freemason – Knight of Malta
George W. Bush – Skull and Bones – Knight of Malta

World Leaders (since 1900)

Tony Blair – 33° Freemason – Knight of Malta – Prime Minister of England
Sir Winston Churchill – 33° Freemason – Prime Minister of England
Josef Stalin – 33° Freemason – Leader of the Soviet Union
Juan Perón – 33° Freemason – President of Argentina
Giscard d'Estaing – Knight of Malta – President of France
Nelson Mandela – Knight of Malta – President South Africa
Juan Carlos – Knight of Malta – King of Spain
Augusto Pinochet – Knight of Malta – President of Chile
Saddam Hussein – 33° Freemason – President of Iraq
John G. Diefenbaker – Freemason – Prime Minister of Canada 1957-1963
Otto von Hapsburg – Knight of Malta – Crown Prince of Austria-Hungary
Bob Hawke – Freemason – Prime Minister of Australia
King Hussein- 33° Freemason – King of Jordan

Intelligence and Military (since 1900)

Allen Dulles – 33° Freemason – Knight of Malta – OSS – CIA – head of MJ-12 – Operation Mockingbird – MK Ultra – head of JFK Warren Commission
James Jesus Angelton – Knight of Malta – CIA counter-intelligence chief
General Reinhard Gehlen – Knight of Malta – head of Nazi intelligence
Heinrich Himmler – Knight of Malta – head of Nazi SS
J. Edgar Hoover – 33° Freemason – Knight of Malta – MJ-12 – head of FBI 1932-1972
General William "Wild Bill" Donovan – Knight of Malta – OSS
Robert McNamara – 33° Freemason – Secretary of Defense – Gulf of Tonkin False Flag
William Casey – Knight of Malta – CIA Director
General Colin L. Powell – 33° Freemason – Secretary of State for George W. Bush
Admiral Richard E. Byrd – 33° Freemason – Operation High Jump
General Douglas MacArthur – 33° Freemason – Interplanetary Phenomena Unit
Frank C Carlucci – Knight of Malta – Secretary of Defense – Deputy Director CIA and National Security Advisor
Oliver North – Knight of Malta – National Security Council staff during the Iran–Contra affair
Francis L. Kellogg – Knight of Malta – CIA – Assistant to Henry Kissinger
George J. Tenet – Knight of Malta – Director CIA
Leon Panetta – Knight of Malta – Secretary of Defense – Director CIA

Historical and other Political Key Figures

David Rockefeller – Knight of Malta – CEO Chase Manhattan Corporation – Trilateral Commission
Zibignew Brezezinski – Knight of Malta – National Security Advisor – Trilateral Commission
McGeorge Bundy – Skull and Bones – National Security Advisor
Henry Kissinger – 33° Freemason – Knight of Malta – National Security Advisor
John Wilkes Booth 33° Freemason – Assassin of President Abraham Lincoln
Paul Warburg – 33° Freemason – Federal Reserve act of 1913
John Dulles – 33° Freemason – Secretary of State – brother of 33° Allen Dulles
John Kerry – Skull and Bones – Secretary of State

Prescott Bush – Skull and Bones – Charged with trading with the enemy (Nazis)
Supreme Court Justice Earl Warren – 33° Freemason – lead JFK Warren Commission
Werner von Braun – 33° Freemason – Project Paperclip Nazi Scientist with NASA (found only one reference to this)
Pat Buchanan – Knight of Malta – Senior advisor to Nixon, Ford and Reagan
Licio Gelli – Knight of Malta – Freemason – Third Reich in Italy initiated Juan Perón into Freemasonry
Ted Kennedy – Knight of Malta – US Senator
Joseph P. Kennedy, Sr.- Knight of Malta – US ambassador to the UK – Chairman of SEC
Thomas 'Tip' O'Neill – Knight of Malta – Speaker House of Representatives
Rick Santorum – Knight of Malta – US Senator
Henry Ford – 33° Freemason – Vehicle manufacturer that supplied for the Nazis
Vladimir Lenin – 33° Freemason – Russian communist revolutionary 1922-24
Karl Marx – 33° Freemason – Russian socialist revolutionary
Leon Trotsky – 33° Freemason – Marxist revolutionary leader of the Bolsheviks
Albert Pike – 33° Freemason – Captain Confederate Army – Wrote Morals & Dogma
Prince Phillip – 33° Freemason – Husband of Queen Elizabeth II
Jeb Bush – Knight of Malta – Governer of Florida
H.G. Wells – 33° Freemason – Science fiction writer of The War of the Worlds
Amschel Mayer von Rothschild – Knight of Malta – Architect of the Bavarian Illuminati
Billy Graham – 33° Freemason – A major influence to Evangelical Christians
Reverend Jesse Jackson – 33° Freemason
Oral Roberts – 33° Freemason – Religious Leader
Joseph Smith – Freemason – Founder of Mormon Church
Jesse Helms – 33° Freemason
Jack Kemp – 33° Freemason
Al Gore – Freemason
Barry Goldwater – 33° Freemason
Newt Gingrich – 33° Freemason
Storm Thurmond – 33° Freemason
Michael Bloomberg – Knight of Malta – Mayor of New York
Michael Chertoff – Knight of Malta – Secretary of Homeland Security
Rudy Giuliani – Knight of Malta – Mayor of New York

Media and Entertainment

Walt Disney – 33° Freemason – Walt Disney Studios
Gene Roddenberry – 33° Freemason – Creator of Star Trek
Darryl Zanuck – Freemason – 20th Century Fox production chief (produced the Day the Earth stood Still)
Jack Warner – Freemason – Warner Brothers Studios Hollywood
Carl Laemmle – Freemason – Universal Studios
Cecil B. deMille – Freemason – Hollywood movie director
Louis B. Mayer – Freemason – Metro-Goldwyn-Mayer
Walter Cronkite – Freemason – Newscaster
Director Ron Howard – 33° Freemason – Apollo 13
Rupert Murdoch – Knight of Malta – head of largest media corporations
William F. Buckley, Jr. – Knight of Malta – Skull and Bones – CIA – TV personality and commentator
Henry Luce – Knight of Malta – Magazine Magnate of Time, Life, Fortune etc.
William Randolph Hearst – Knight of Malta – Newspaper Magnate
Pat Buchanan – Knight of Malta – CNN political analyst
John Wayne – Freemason – Actor
Clark Gable – Freemason – Actor
Will Smith – Freemason – Actor (note this is a very partial actor list)

Jesuits in the White House Now and Then

> **The Jesuit Oath**
> *The Counter-Reformation War*
>
> I furthermore promise and declare that I will, when opportunity present, make and wage relentless war, secretly or openly, against all heretics, Protestants and Liberals, as I am directed to do, to extirpate and exterminate them from the face of the whole earth; and that I will spare neither age, sex or condition; and that I will hang, waste, boil, flay, strangle and bury alive these infamous heretics, rip up the stomachs and wombs of their women and crush their infants' heads against the walls, in order to annihilate forever their execrable race. That when the same cannot be done openly, I will secretly use the poisoned cup, the strangulating cord, the steel of the poniard or the leaden bullet, regardless of the honor, rank dignity, or authority of the person or persons, whatever may be their condition in life, either public or private, as I at any time may be directed so to do by any agent of the Pope or Superior of the Brotherhood of the Holy Faith, of the Society of Jesus.
>
> http://uncontrolledopposition.com - Find us on Facebook

US President Donald Trump Family and Administration

Donald John Trump, Jr. – Alma mater: University of Pennsylvania
Ivanka Marie Trump – Alma mater: University of Pennsylvania
Eric Frederic Trump – Alma mater: Georgetown University; member of "Business Society and Public Policy Initiative Board of Advisors"

Mike Pence, Vice-President - Georgetown University in 1992. He attends the Vatican's "Red Mass" and is a recipient of the Roman Catholic "John Carroll Award."
Mike Pompeo, Director of CIA - Pompeo; his wife, Susan; and their son, Nicholas, reportedly attend Eastminster Presbyterian Church, where he serves as a deacon and has taught the fifth-grade Sunday school class.
Steven Bannon, Chief of Staff - **Master's Degree in national security studies at Georgetown University;** Harvard Business School; Senior Investment Banker, **Goldman Sachs NYC.**
Betsy DeVos, Secretary of Education. Sister of Erik Prince, founder of Blackwater/Academi/Xe private mercenary army. Graduated from Calvin College and attending Mars Hill Bible Church in Grand Rapids, Michigan.
Jeff Sessions, US Attorney General, Huntingdon College in Montgomery, affiliated with the United Methodist Church, active in his family's church, Ashland Place United Methodist Church in Mobile, where he's served as a lay leader, Sunday school teacher and chairman of its administrative board. He also has been selected as a delegate to the annual Alabama United Methodist Conference.
Sean Spicer, President Speech Writer - He has been the Republican National Committee's communications director since 2011 and a chief strategist since 2015. He also worked as a senior communications adviser for Trump during the transition. Spicer graduated in 1989 from Portsmouth

Abbey School, a Benedictine boarding and day school in Portsmouth, Rhode Island, before going on to graduate from Connecticut College and receive a master's degree in national security and strategic studies from the Naval War College in Newport, Rhode Island.

Wilbur Ross, Secretary of Commerce – Attended Jesuit Xavier Catholic School. Senior Director of Rothschild Inc.

Andrew Bremberg - Director of the White House Domestic Policy Council; Attended Franciscan University of Steubenville.

Tom Price, Head the Department of Health and Human Services Price is a Presbyterian who attended Emory University, affiliated with the United Methodist Church. In Congress, Price's legislation and voting record reflect an alignment with conservative Christianity.

Reince Priebus , GOP Chairman - Priebus has worked with Archbishop Timothy Dolan, the Catholic Archbishop of New York, to help change both the Republican party and the face of the Church to focus on more serious topics such as abortion, pre-marital sex, homosexuality, same-sex marriage, birth control, stem cells and the ordination of women.

David Malpass - Jesuit-trained from Georgetown, Vice President of the Council for National Policy, leading appointment selections for positions involving economic issues

Keith Kellogg - trained by Jesuit at Santa Clara University, leading appointment selections for positions involving national defense issues

Michael Catanzaro - trained by Jesuits at Fordham University and St. Ignatius High School, leading the policy implementation team for energy independence

Andrew Bremberg - graduate of Catholic University of America Executive Legal Action Lead

James Carafano - Jesuit-trained from Georgetown University , reported to be the primary aide to the State Department of Trump administration transition team

Ed Feulner - Roman Catholic former President and founder of Heritage Foundation; Jesuit-trained from Regis and Georgetown Universities

Ken Blackwell - Jesuit-trained from Xavier University, leading appointment selections for positions involving domestic issues.

Boris Epshteyn - Trump's foremost spokesman; Jesuit-trained from Georgetown.

"Between 1555 and 1931 the Society of Jesus [i.e., the Jesuit Order] was expelled from at least 83 countries, city states and cities, for engaging in political intrigue and subversion plots against the welfare of the State, according to the records of a Jesuit priest of repute [i.e., Thomas J. Campbell]. Practically every instance of expulsion was for political intrigue, political infiltration, political subversion, and inciting to political insurrection." (1987) J.E.C. Shepherd (Canadian historian)

The Military Generals in the Trump Administration

General James Mattis, Secretary of Defense – Mattis popularized the 1st Marine Division's motto "no better friend, no worse enemy," a paraphrase of the famous self-made epitaph for the Roman dictator Lucius Cornelius Sulla,[25] in his open letter to all men within the division for their return to Iraq.

General John Kelly, Department of Homeland Security - Boston native raised in an Irish-Catholic family, Kelly served in the Marines and led the invasion into Iraq before leading up the U.S. Southern Command. Previously he ran Guantanamo Bay prison.

General Vincent Viola, Secretary of Army – nicknamed the "warrior monk". a businessman who founded a high-frequency trading firm, Virtu Financial, and is also owner of the Florida Panthers. He has deep ties to Fordham University, where, in 2009, he donated $2 million to endow a chair named after Jesuit Cardinal Avery Dulles. "In all of those roles, he has done much to advance Fordham's Jesuit, Catholic mission and strategic priorities, and we are deeply in his debt," said Jesuit Father Joseph McShane, the president of Fordham.

US President Barrack Obama's Administration

Joseph Biden, Vice-President, St. Joseph's University
Dan Pfeiffer, Deputy Communications Director, St. Joseph's University.
John Brennan, CIA Director, Jesuit Fordham University
Robert Cardillo, Director of National Intelligence, Georgetown
Tom Donilon, Deputy National Secretary, Catholic University of America
Rodney Snider, Sr. Director for Intelligence Program, Georgetown University.
Robert Gates, Secretary of Defense, Georgetown University
Jacob Lew, Office of Management and Budget Director, Georgetown University
William Daley, White House Chief of Staff, Loyola University Chicago

General James L. Jones, National Security Advisor, Georgetown University.
General David Petraeus, Southern Commander of Iraq Invasion, Georgetown University
General John Allen, Deputy Commander of US Central Command, Georgetown University.

Congress/Governors/Mayors/State Departments; to name just a few

Bill Clinton, 42nd President, Georgetown University
Chris Christie, Governor of New Jersey, "I'm a Catholic, but I've used birth control, and not just the rhythm method," Christie declared. Christie's wife, who grew up in a large Catholic family, taught religious courses at their home parish, St. Joseph Church in Mendham, New Jersey. His children attend Catholic school.
Newt Gingrich, Senator, member of Council on Foreign Relations Mr. Gingrich decided to become a Roman Catholic after having been born a Lutheran and joining the Southern Baptist Church in college. In 2009, he was baptized in the same Catholic parish church on Capitol Hill where Senator Robert F. Kennedy once attended noonday Mass and sometimes assisted the priest as an altar server.
Rudy Giuliani, NY Mayor, 9/11 coadjutor, Knight of Malta. He was raised a Roman Catholic. He attended the local Catholic school, St. Anne's. Later, he commuted back to Brooklyn to attend Bishop Loughlin Memorial High School, graduating in 1961.
Janet Napolitano, Secretary of Homeland Security (Obama Administration) and former Governor of Arizona; President of University of California Regents. Technology school of the Jesuits, Santa Clara University
Jerry Brown, Former and current (2011) Governor of California; former Mayor of Oakland, California. Santa Clara University
Gavin Newsom, Lt. Governor of California, Mayor of San Francisco, Santa Clara University.
Leon Panetta, CIA Director, Chief of Staff to Bill Clinton, Fordham University.
William Casey, CIA Director, Fordham University.
Dwight D. Eisenhower, US President, Fordham University

Resources/Books

Books *(available in .pdf)*

Earth Not A Globe!	Samuel Rowbotham (Parallax)	1865/01/01
Kings Dethroned	Gerard Hickson	1922-01-01
Zetetic Astronomy	Lady Blount, Albert Smith	1904-01-01
Is Newtonian Astronomy True?	William Carpenter	1895/05/01
Heaven and Earth	Gabrielle Henriete	1956-01-01
Flat Earth Conspiracy	Eric Dubay	2015/11/09
Terra Firma	David Wardlaw Scott	

Trusted Flat Earth Subject Websites:

Aplanetruth.info
Whotfetw.com (extensive daily news and disclosures)
FlatEarthperspectives.com
Paulmichaelbales.com
Testingtheglobe.com
Flatearth.click

You Tube Channels:
Aplanetruth.info
Brian Mullins
Oddtv
Planate Veritas
Rob Skiba
The Morgille

Links:

Nasa Future of War
http://www.stopthecrime.net/docs/nasa-thefutureof-war.pdf
Silent Weapons for Quiet Wars
http://www.stopthecrime.net/docs/SILENT%20WEAPONS%20for%20QUIET%20WARS.pdf
Owning the Weather by 2025
http://csat.au.af.mil/2025/volume3/vol3ch15.pdf
Report from Iron Mountain
http://stopthecrime.net/docs/Report_from_Iron_Mountain.pdf
Geoengineering and Weather Manipulation
http://www.geoengineeringwatch.org/

Other Author Books

**Geoengineering aka Chemtrails
Investigations into Humanity's
6th Great Extinction Event**

**Everything is a Lie
Nothing is True
Critical Subjects Few Know Anything About
But Should**

Websites
www.tabublog.com
www.aplanetruth.info
www.avvi.info
You Tube Channel:
aplanetruth.info

All Health Store   TheDWDGshop

 James (Jamie) Lee resides near the Mendocino Coast of Northern California growing biodynamic and organic food at his 100-yr. old farm.

He graduated from the school of business at San Diego State University as well as attended the Green MBA program at New College in Santa Rosa, California.

He has had over 25 years experience working on Wall Street beginning working for the investment banking firms, Furman, Selz, Inc. in New York City as an Institutional Sales Trader before moving back to San Francisco, California to work for Robertson, Colman, Stephens Investment Bank. In 1991, he founded a small investment/research boutique, JWL Investments.

Using his Wall Street experience for the past 10 years and spending 6 – 8 hours a day for years… analyzing, investigating and exploring the hidden occult world behind the power bases, Jamie brings the reader up to date of why nothing gets better and what is planned for the human 'resources' by the wealthy religious elite.

His work has been published on many alternative news websites including Waking Times, Activist Post, Philosophers-Stone, Reddit, Beforeitsnews.com, David Icke, Rense.com, Sage of Quay and 2015 Most Censored Stories as well as appeared on the internationally syndicated evening news show, Breaking the Set with Abby Martin in November of 2014 about legislation passed in Mendocino County, California declaring local rights of self-governance and determination preempting state, federal and international law.

"Inaction deemed more deadly than error." — David Davidson

Printed in Great Britain
by Amazon

73277447R00100